Hume's Difficulty

Donald L. M. Baxter's meticulous attention to textual detail yields a highly original interpretation of some of the most neglected or maligned parts of Hume's *Treatise*. The focus is Hume's treatment of the concept of numerical identity, which is central to his famous discussions of the external world and personal identity. Hume raises a long unappreciated, and still unresolved, difficulty with the concept of identity: How to represent something as "a medium betwixt unity and number." Superficial resemblance to Frege's famous puzzle has kept the difficulty in the shadows. Hume's way of addressing it makes sense only in the context of his unorthodox theory of time. Baxter shows the defensibility of that theory against past dismissive interpretations, especially of Hume's stance on infinite divisibility. Later the author shows how the difficulty underlies Hume's later worries about his theory of personal identity, in a new reading motivated by Hume's important appeals to consciousness. Baxter casts Hume throughout as an acute metaphysician and reconciles this side of Hume with his overarching Pyrrhonian skepticism.

The book will be useful to those interested in the metaphysics of identity and time and the epistemology of metaphysics and will be indispensable to Hume scholars, who have lacked an in-depth treatment of these crucial and intricate issues. It is the first focused study of Hume on time and identity.

Donald L. M. Baxter is Professor of Philosophy at the University of Connecticut.

Routledge Studies in Eighteenth Century Philosophy

1 Naturalization of the Soul
Self and personal identity in the eighteenth century
Raymond Martin and John Barresi

2 Hume's Aesthetic Theory
Taste and sentiment
Dabney Townsend

3 Thomas Reid and Scepticism
His reliabilist response
Philip de Bary

4 Hume's Philosophy of the Self
A. E. Pitson

5 Hume, Reason and Morality
A legacy of contradiction
Sophie Botros

6 Aesthetics and Morals in the Philosophy of David Hume
Timothy M. Costello

Hume's Difficulty

Time and identity in the *Treatise*

Donald L. M. Baxter

Routledge
Taylor & Francis Group
LONDON AND NEW YORK

First published 2008
by Routledge
2 Park Square, Milton Park, Abingdon, Oxon, OX14 4RN

Simultaneously published in the UK
by Routledge
270 Madison Ave, New York NY 10016

Routledge is an imprint of the Taylor & Francis Group, an informa business

Transferred to Digital Printing 2009

Typeset in Times New Roman MT by
Taylor & Francis Books

Library of Congress Cataloging in Publication Data
A catalogue record for the book has been requested

British Library Cataloguing in Publication Data
A catalog record for this book is available from the British Library

ISBN10: 0-415-95594-7 (hbk)
ISBN10: 0-415-80477-9 (pbk)
ISBN10: 0-203-94025-3 (ebk)

ISBN13: 978-0-415-95594-2 (hbk)
ISBN13: 978-0-415-80477-6 (pbk)
ISBN13: 978-0-203-94025-9 (ebk)

To Robert and Mary, Sandra, and Nicolina Joy

Contents

Acknowledgments

I am grateful to my colleagues and students in philosophy at the University of Connecticut, who have listened and responded to much of the material in this book with the department's characteristic blend of amiability and exacting standards. I have appreciated the opportunities for scholarly interchange offered by the Hume Society. Although notes in each chapter list many, though likely not all, who have helped in that chapter's genesis, I would like here particularly to thank three people: John Troyer, who has read or heard it all with interest and insight; David Fate Norton, for valuable comments and guidance; and Joseph L. Camp, Jr., who helped an ancestor of this and other projects get underway.

The section of Chapter 1 defending the view that Hume is a Pyrrhonian Empiricist is a revision of a section of my "Hume's Theory of Space and Time in its Sceptical Context," forthcoming in David Fate Norton, ed., *The Cambridge Companion to Hume*, 2nd edn (Cambridge: Cambridge University Press).

The four-line argument with premises (1) through (3) in Chapter 2 is an improved version of an argument presented in my "Hume on Infinite Divisibility," *History of Philosophy Quarterly* 5 (1988): 133–40. The supporting material, while in the same spirit, is different.

Chapter 3 is a revision of my "Hume on Steadfast Objects and Time," *Hume Studies* 27 (2001): 129–48, which was a major revision and considerable expansion of my treatment of Hume on time in "A Defense of Hume on Identity through Time," *Hume Studies* 13 (1987): 323–42. An earlier version of parts 2 and 3 of the 2001 paper appeared as "A Humean Temporal Logic," in *Proceedings of the Twentieth World Congress of Philosophy, Volume VI: Analytic Philosophy and Logic*, ed. Akihiro Kanamori (Bowling Green, Ohio: Philosophy Documentation Center, 2000), 209–16.

The discussion of Green and Stroud toward the end of Chapter 4 is drawn from a couple of pages of "A Defense of Hume on Identity through Time," *Hume Studies* 13 (1987): 323–42.

Chapter 5 began as a revision of "Hume's Labyrinth Concerning the Idea of Personal Identity," *Hume Studies* 24 (1998): 203–33, but became an entirely different essay. Only a couple of pages of discussion of competing interpretations have survived. Acknowledgment is made to *Hume Studies*, in which earlier versions of substantial parts of several chapters first appeared.

Chapter 6 is a revision of "Hume's Puzzle About Identity," *Philosophical Studies* 98 (2000): 187–201. Copyright 2000 Kluwer Academic Publishers. With kind permission of Springer Science and Business Media.

At various stages, Christopher Panza, Donovan Cox, Adam Potthast, and Patrick Fleming provided helpful research assistance.

Abbreviations

Hume texts

Dialogues	Norman Kemp Smith (ed.) *Hume's Dialogues Concerning Natural Religion* (Oxford: Clarendon Press, 1935).
EHU	*An Enquiry Concerning Human Understanding*, edited by Tom L. Beauchamp (Oxford: Clarendon Press, 2000), cited by section.paragraph.
History	*The History of England, from the Invasion of Julius Caesar to the Revolution in 1688*, 6 vols (Indianapolis, Ind.: Liberty Classics, 1983).
OST	"Of the Standard of Taste," in *Essays Moral, Political, and Literary*, edited by Eugene F. Miller (Indianapolis, Ind.: Liberty Fund, 1985).
SBN	*A Treatise of Human Nature*, 2nd edn, edited by L. A. Selby-Bigge and P. H. Nidditch (Oxford: Clarendon Press, 1978).
SBN	*Enquiries Concerning Human Understanding and Concerning the Principles of Morals*, 3rd edn, edited by L. A. Selby-Bigge and P. H. Nidditch (Oxford: Clarendon Press, 1975).
T	*A Treatise of Human Nature*, edited by David Fate Norton and Mary J. Norton (Oxford: Oxford University Press, 2000), cited by book.part.section.paragraph.

Also specially cited

Essay	John Locke, *An Essay Concerning Human Understanding*, edited by Peter H. Nidditch (Oxford: Clarendon Press, 1975), cited by book.chapter.section.

Introduction

David Hume's account of identity—numerical identity—is central to his *Treatise* Book 1 project of explaining how we come to believe in unitary selves and a world of internal perceptions and external objects, variously united by relations of resemblance, space, time, and causation. According to Hume, we need an idea of identity in order to come to believe in external objects, and in the self. Yet there has been no in-depth study of Hume's view of identity, no full explanation of what he means when he takes it to be "a medium betwixt unity and number." At best there is paraphrase of his view, as is found in Kemp Smith, or assimilation to Frege's view of identity sentences, as is found in Bennett.[1]

Understanding Hume's account of identity requires understanding his account of time; he links these two important concepts. Especially required is an understanding of his contrast between things with duration, namely successions, and things that coexist with successions yet lack duration, namely steadfast objects. Yet there has been no in-depth study of Hume's account of time. At best there has been dismissive critique of aspects of it, as is found for instance in Flew.[2]

Insofar as Hume is giving accounts of time and identity, he is doing metaphysics. There have been no attempts to reconcile Hume's essays into metaphysics with his thoroughgoing skepticism. His rejection of "school metaphysics" in the first *Enquiry* may have misled many into thinking he rejects metaphysics in general. He does not. In the same way that he accepts any part of his science of man in the face of unanswerable skeptical challenges, he accepts some metaphysics.

The lack of appreciation of Hume's account of identity, has meant that no one has realized that at the basis of his account is a problem in the concept of identity for which we still have no solution. There is a need to understand Hume on identity for systematic reasons, as well as scholarly and historical ones.

The present work is meant to supply these deficiencies by being the first focused study of Hume on time and identity. By close reading, careful reconstruction, and argument sensitive to the concerns of historians of philosophy and contemporary metaphysicians, I detail his unusual views, show their surprising strengths, and argue that he is right that there is a difficulty concerning the concept of numerical identity.

The central concern in Hume's account of identity is to capture how it can be that we can think of some things while leaving open whether or not they are identical—as we do anytime we are unsure whether some things we are acquainted with are identical. Somehow we are able to consider both alternatives: that they are and that they are not. However, if we think of them as one and the same, we seemingly cannot conceive how that thing could possibly be distinct from itself. If we think of them as two and distinct, we cannot conceive how those two could possibly be identical. But what is the alternative to thinking of them as one thing or thinking of them as more than one thing? Somehow we have to be able to take two views of the same things such that they are one and the same on one view, and two distinct things on the other view. "They" can't be simply one thing, nor simply many things, but must somehow be "a medium betwixt unity and number."

The only way Hume can think of to approximate this "medium" is in cases of identity through time. Such cases involve the fiction of a steadfast object with duration. According to Hume's account of time, something with duration is really many distinct things in succession. Something steadfast, on the other hand, is a single unitary thing coexisting with a succession. So the fiction of a steadfast object with duration is the fiction of something that is one single thing and yet many distinct things—just what Hume needs to be a medium betwixt unity and number. I explain and motivate Hume's unorthodox account of time, and show how he tries to use it to meet the needs of his account of identity.

It turns out, however, that the central concern remains an unresolved difficulty, and not just for Hume. I show that the difficulty can be separated from Humean peculiarities, is genuine, and has not yet been adequately addressed.

Let me give a bit more detail. Hume says that thinking of things *a* and *b* as identical requires being able to think of them as one yet also as many, and so requires thinking inconsistently. He seems right. If we are unsure of their identity, then as we contemplate the alternatives we must be able to think of *a* and *b* as one and the same, but also be able to think of them, the same them, as two distinct things. But this means that when we are thinking of them as one thing, we must be able to go on to think of it (that one thing) as distinct from itself. After all, if we can think of each of them as it, then we can think of it as each of them, and so as each of the two distinct things. So there seems to be a contradiction at the heart of identity. To see the contradiction we have to focus on the fact that it is *a* and *b* that we are thinking about. It is them that we have to hold fixed when considering the alternatives, not their names nor descriptions nor evidences of them. Further, we are thinking of them alternately as identical and distinct from each other, not alternately as composing a single whole and being its many parts.

Seeing the difficulty requires distinguishing (i) what there is which a representation represents, from (ii) what a representation represents there as being. These awkward locutions better capture a distinction like that made by Ayers between "real object" and "intentional object of thought."[3] Hume is not so much concerned with how many things there actually are in the world, that we are thinking about when wondering about an identity. Rather, he is concerned with how many things we represent there as being, when we wonder. For example, Cicero and Tully may actually be one and the same person in the world. However, it cannot be that we represent them as being one and the same when wondering about their identity, for then it would seem impossible to think that they might be distinct.

Chapter 6 is a systematic attempt to show that Hume's difficulty is a genuine problem. Chapter 4 argues that this is the problem Hume posed, and thought he had satisfactorily evaded, in his discussion of the idea of identity. He characterizes it as an idea of a "medium betwixt unity and number," or better, "as either of them, according to the view, in which we take it" (T 1.4.2.29; SBN 201). In order to explain how we form this two-faced idea, Hume supposes that identity is just identity through time. It is only in coming to think of things in time, he contends, that we come to think of something that is one single thing viewed one way, and many distinct things viewed another. However, the idea is, perforce, a "fiction of the imagination" that masks contradiction, because "[b]etwixt unity and number there can be no medium" (T 1.4.2.28; SBN 200). Chapter 5 argues that Hume would need to solve the difficulty about identity in order to resolve his worry about his account of *personal* identity. In the Appendix to the *Treatise* he claims to have found an inconsistency in his account of how consciousness comes to regard many successive past perceptions as one and the same self. I argue that, given his commitment to its accuracy, consciousness could take the distinct perceptions to be identical only if they were somehow distinct in one way and yet identical in another. But to make sense of this situation one would have to be able to make sense of "a medium betwixt unity and number."

Hume thinks that one must appeal to time to address the difficulty with identity. Thus, understanding Hume on identity requires understanding his account of time. Chapter 2 explains that there are temporal simples, which are *single* things, and that something with duration is really *many* temporal simples in succession. In order to establish that there are simples, Hume is at pains to show that not every part of something has parts. His arguments against infinite divisibility in this sense, are the relevant ones. The result about duration follows from Hume's arguments that duration is merely an abstraction from successions, i.e., their successiveness. Chapter 3 explains that some temporal simples ("stedfast objects") coexist with successions. Thus, despite being a single thing lacking duration, a steadfast object coexists with a many that has duration. This last, strange view is crucial to understanding Hume on the idea of identity, which results from the natural fiction that a steadfast object does have duration. It is not that the idea of a steadfast object is itself an idea of identity; it is rather that the idea of a steadfast object both as such and yet additionally as having duration, is the idea of identity. The idea of the steadfast object as such is an idea of unity; the fictional idea of the steadfast object as having duration is an idea of number; and the idea of identity is composed of these two views of the steadfast object. Thus the object is somehow represented both as only one thing and yet as many things.

Hume thinks he explains the "somehow" in a way suitable for his purposes, but it turns out he does not. Nor do we know how to, either, even for our own purposes.

I have Hume engaged in metaphysics, which seems incongruous for a self-professed skeptic. On what basis can a skeptic claim to know anything about, or even to have reasonable beliefs about, time or identity? But ultimately Hume doesn't claim to know anything or to have any reasonable beliefs. As I argue in Chapter 1, he just gives vent to the views that strike him most forcibly, including fundamental metaphysical presuppositions supported by weighty tradition or by his own well-developed metaphysical taste. As for the rarified sort of assent that is supposed simply to be the rational result of having good reasons, that assent he withholds. He never finds sufficiently good reasons, and it appears to him that no one ever could. In assenting and not assenting in these differing ways, Hume is to that extent a Pyrhonnian skeptic, though perhaps without fully realizing it.

In giving my interpretation, I am more interested in close reading of the text and getting everything to hang together than in making Hume agree with our own views of the issues he discusses. The latter assumes that we have things to teach Hume; the former assumes that he has things to teach us.

If philosophy is dialogue, then doing history of philosophy is mainly a matter of listening. To listen best, one must try critically to believe what the other is saying. The uncritical believer and the critical disbeliever both miss more of what is said. In interpreting what the other is saying, one must try to make all of it as defensible as possible, which is different from making it conform as much as possible to what one holds to be true. Presumably one can regard as defensible theories one takes to be almost entirely false. Thus the maxim I am recommending is not the Principle of Charity; it should perhaps be called the Principle of Defensibility.[4] The rule of thumb is: If an interpretation has a great philosopher saying something indefensible, then it is better to revise the interpretation than to reject the philosopher's claim. The resulting reading preserves everything the philosopher says but often contains details beyond what the philosopher explicitly provides. It is like a more detailed mechanical drawing of a machine sketched by Leonardo that shows more precisely how the machine would work. So in interpreting Hume, I will supply arguments or views that, if assumed to underlie what he says, explain and defend what he says.

Of course, after one has listened with respect, it may well turn out complete belief cannot and should not be sustained. Still, all but central obstacles to belief will have been overcome. As a result, one will to one's best ability have appreciated some of the other's main insights. The great thinkers show us what is deepest and most important in philosophy. Their concerns are not just historical curiosities or footnotes to current debates. Their concerns are rather at the wellsprings of philosophy and we should let them lead us there.

1 Interpreting Hume as metaphysician and skeptic

Metaphysician

On my interpretation, David Hume is a great metaphysician. He systematically applies an elegant metaphysics based on a few deep assumptions. Among them are:

1. Only single things exist (T 1.2.2.3; SBN 30).
2. Only particular things exist (T 1.1.7.6; SBN 19).
3. Alteration is contrary to identity (T 1.4.3.2; SBN 219).
4. Interruption is contrary to identity (T 1.4.2.24; SBN 199).
5. Anything divisible has parts (T 1.2.2.2; SBN 29).[1]
6. Composition is contrary to unity (T 1.2.2.3; SBN 31).
7. Things are mentally separable if and only if numerically distinct (T 1.1.7.3; SBN 18).
8. The conceivable is possible (T 1.1.7.6; SBN 19–20).
9. There is no middle way between existing and not existing (T 1.4.2.28; SBN 200).

Based on these assumptions and others, Hume presupposes and sketches a metaphysics of space and time, of parts and whole, of identity and personal identity, among other things. Metaphysics was not his main concern, of course. He was, rather, concerned to make deeper and more consistent the skeptical empiricism of his time and to apply it to the study of the mind. Along the way, however, he made contributions to metaphysics as well.

Ranking Hume as a great metaphysician may seem to conflict with his injunction at the end of the first *Enquiry* to burn any works of metaphysics. Note, however, that he there is specifically talking of an example of "divinity or school metaphysics." Clearly he has a restricted class of works in metaphysics in mind. The *Enquiry* itself, after all, is a work in "what is commonly called *metaphysics*" (EHU 1.7; SBN 9). In other words, it is a case of the "profound" or "abstruse" reasoning he takes such care to defend in the first section. He is hardly recommending burning the *Enquiry* along with the others.[2]

Still, the worry about Hume and metaphysics is not settled so easily. As I am using the word, it means less than simply abstruse reasoning and more than just school metaphysics. Metaphysics in my construal is an attempt to systematize our preconceptions involving fundamental concepts such as existence, identity, number, time, causation, and others. Although the starting points are natural or commonsense beliefs, some of them might be sacrificed or revised for the sake of the systematization. Hume's *Enquiry*, to the extent that its main concerns are with clarifying our ideas and proportioning our belief to the evidence, falls outside of metaphysics in this intermediate sense. And it does appear that metaphysics in this sense does not meet Hume's test. Metaphysics is not mathematics on the one hand, nor is it empirical science on the other. So Hume the great metaphysician still seems opposed by Hume the book-burner.

Perhaps metaphysics as in the *Treatise* was a youthful folly that a maturer and severer Hume condemned. But no. In the *Enquiry* he still endorses his metaphysical reasonings about infinite divisibility from the *Treatise* (EHU 12.18, note 33; SBN 156, note 1). What then is he condemning?

He is condemning books that purport to be contributions to knowledge that have *not any* mathematical or experimental reasoning—none. This is not to endorse only books that are *all* mathematical or experimental reasoning; it is just to endorse only those with *some*, such as his own. A reasonable interpretation is that Hume is condemning books whose central claims are inconsistent with the results of proper mathematical or experimental reasoning. It would be folly for him to ask for more. Hume's own conclusion about the sort of book he rejects—"it can contain nothing but sophistry and illusion"—is a conclusion that neither science nor mathematics can establish (EHU 12.34; SBN 165).[3] Just insofar as he assumes that there are numerically distinct objects in space and time, as he does in his discussion of causation, he is making metaphysical assumptions that science presupposes and mathematics is silent about. No book about the world can contain no metaphysical assumptions.

Further, the book-burning image does not reflect Hume's considered view given his plea for tolerance in speculation (EHU 11.1–9; SBN 132–5). He is likely using the image to remind his readers of the consequences of dogmatic superstition; to remind his readers of the books and, horribly, the people burned by those he opposes; to remind his readers that the stakes are high.

There is no conflict, then, between Hume the mathematical and experimental reasoner and Hume the metaphysician. As long as his metaphysics is consistent with his empiricism, it is innocent enough.

Skeptic

His metaphysics may be consistent with his empiricism, but how can it be consistent with his skepticism? How could someone employ metaphysical principles if, as skeptic, he finds no reason to believe them?

The problem is similar to the familiar problem of reconciling Hume's stance as a skeptic undermining all belief, with his stance as an empirical mental naturalist giving a constructive account of the workings of the mind. Giving a more detailed reconciliation of these two Humes will make room for Hume the metaphysician as well. What is needed is a fuller understanding of his skepticism.

Viewing skepticism selectively with the narrow epistemological focus characteristic of early modern philosophy, then Hume is a Pyrrhonian skeptic who emulates an Academic mitigated skeptic. To put it more precisely, Hume adapts the Pyrrhonian stance on belief and justification in a way that allows him to emulate Academic mitigated skeptics in matters beyond active daily life. Whether by accident or design, he is Pyrrhonian more in the way described by Sextus Empiricus than in the way Hume himself describes. Hume takes Pyrrhonists to assent to nothing in any way, to have simply "no opinion nor principle concerning any subject, either of action or speculation" (EHU 12.2; SBN 149). Perhaps Hume's interpretation of Pyrrhonism was somewhat distorted by the commentators he was familiar with, or perhaps he found it useful just to appeal to a widely extant interpretation.[4] In any event, to understand the complexities of Hume's skepticism, it helps to disregard his own characterizations and to apply the nuanced descriptions of ancient skepticism given by the ancient sources themselves and some exacting commentators of our own day.[5] Such descriptions give us some key features to look for in a skeptic's position which we might otherwise overlook.

The Pyrrhonians distinguished two kinds of assent: (i) active endorsement of a view as true based on an appropriate reason, and (ii) passive acquiescence in a view forced upon one by appearances.[6] In seeking after truth, they found that any reasons for endorsing a view as true could be counterbalanced by reasons for not so endorsing it. They found themselves suspended, unable to endorse any view or the opposite of any view. Serendipitously, they found serenity in this state of suspension. This suspension of judgment was not, contrary to the contentions of Hume and many of his predecessors, supposed to be a suspension of all assent whatsoever. The Pyrrhonians, while continuing to seek the truth about reality, allowed themselves, in the course of daily life, to acquiesce in whatever view happened to be forced upon them by the appearances of things. If they appeared to be at the edge of a cliff in danger of falling and if falling appeared to be something that would bring on pain, they would try not to fall. They would do so, however, without endorsing as true these views that were motivating them.

The Academic mitigated skeptics were a bit more dogmatic than the Pyrrhonians. They found that they retained certain beliefs after weighing the conflicting arguments, and regarded this greater "probability"—i.e., plausibility or convincingness—as reasonable evidence (though in no way conclusive) that those beliefs were true. Thus they found it plausible that truth could at least be approached, even if, as seemed very plausible, never apprehended.[7]

The Pyrrhonians undercut philosophy, science, and mathematics. The late Academics, with their reliance on probable beliefs, provided for these endeavors. Hume admired the modern flowering of such an approach in Newton.[8] However, the Academics in their degree of dogmatism were still vulnerable to a Pyrrhonian-type challenge. Hume gives it in "Of Scepticism with Regard to Reason," where he critiques an Academic reliance on the probable. He argues in effect that the Academic cannot justify any degree of belief in any conclusion (T 1.4.1; SBN 180–7). However, Hume invents a Pyrrhonian facsimile of Academic probability.[9] It involves two kinds of assent, and degrees of the passive kind of assent.

Hume's version of active endorsement is an idealized kind of belief assumed by his opponents and by most of us.[10] It is belief in a conclusion based purely on the fact that one has good reasons for it, which one recognizes as good—that is, belief as an act of "*the cogitative part of our natures,*" as a "simple act of thought." In contrast is belief as "*an act of the sensitive ... part of our natures,*" as a "peculiar manner of conception," as "the addition of force and vivacity" (T 1.4.1.8; SBN 183–4). Here is Hume's passive acquiescence. Resulting from experience and habit, it is the only actual way to believe any matters of fact beyond the present evidence of the senses: "Without this quality, by which the mind enlivens some ideas beyond others (which seemingly is so trivial, and so little founded on reason) we cou'd never assent to any argument, nor carry our view beyond those few objects, which are present to our senses" (T 1.4.7.3; SBN 265).

Even our assurance in "all demonstrative sciences" resolves itself into force and vivacity dependent on experience (T 1.4.1.1; SBN 180). So belief as active endorsement must be a philosopher's invention, though Hume doesn't say so. Perhaps force and liveliness are sometimes like a calm passion, known more by their effects than by any feeling, and so are mistaken for the "determinations" of reason by itself (cf. T 2.3.3.8; SBN 417).

In any event, like the Pyrrhonians, Hume finds no final reason actively to endorse any views as true, or even as probable. He is content passively to acquiesce in whatever views are forced on him by appearances: "After the most accurate and exact of my reasonings, I can give no reason why I shou'd assent to it; and feel nothing but a *strong* propensity to consider objects *strongly* in that view, under which they appear to me" (T 1.4.7.3; SBN 265).

In the first *Enquiry* Hume notes "the whimsical condition of mankind, who must act and reason and believe; though they are not able, by their most diligent enquiry, to satisfy themselves concerning the foundation of these operations, or to remove the objections, which may be raised against them" (EHU 12.23; SBN 160). Nothing with any justification mitigates our Pyrrhonian suspension of active assent. That remains entire. What is mitigated is our natural propensity, after extended careful reasoning concerning such foundations, despairingly to allow the reason-subverting conclusion to be forced upon us, that "We have, therefore, no choice left but betwixt a false reason and none at all." Any mitigation of this "deplorable" condition is only because, after "relaxing this bent of mind" we forget the difficulty. "Most fortunately it happens, that since reason is incapable of dispelling these clouds, nature herself suffices to that purpose" (T 1.4.7.7–9; SBN 268–9). First in common life and eventually in philosophy, we yield to passive acquiescence in many various views. That acquiescence is forced by "the strong power of natural instinct" (EHU 12.25; SBN 162). "Nature, by an absolute and uncontroulable necessity has determin'd us to judge as well as to breathe and feel" (T 1.4.1.7; SBN 183). So the mitigated skepticism recommended by the first *Enquiry* is merely a facsimile of the ancient version that took plausibility to be reasonable evidence of truth.[11] Hume finds no reasonable evidence of truth. As he summarizes in the Abstract of the *Treatise*, "Philosophy would render us entirely *Pyrrhonian*, were not nature too strong for it" (T Abs.27; SBN 657). In yielding to nature, I've argued, Hume is Pyrrhonian as well.

This yielding occurs, for Hume, beyond the confines of active daily life, because "'tis almost impossible for the mind of man to rest, like those of beasts, in that narrow circle of objects, which are the subject of daily conversation and action" (T 1.4.7.13; SBN 271). It is especially impossible for some with a certain bent of mind, for "it is with some minds as with some bodies, which, being endowed with vigorous and florid health, require severe exercise, and reap a pleasure from what, to the generality of mankind, may seem burdensome and laborious" (EHU 1.10; SBN 11).[12]

Hume's main innovation, in adapting the Pyrrhonian approach to skepticism in a way that allows passive assent to views even concerning theoretical matters, is to make distinctions between the relative strength or weakness of the forcing of the views on us by appearances. It is by feeling that the mind distinguishes the degree to which an idea is forced upon it. In allowing for what Sextus calls "strong inclination or adherence," Hume differs from ancient Pyrrhonians who, as Sextus says, yield "without adherence."[13] As Hume says, "An idea assented to *feels* different from a fictitious idea, that the fancy alone presents to us: And this different feeling I endeavour to explain by calling it a superior *force*, or *vivacity*, or *solidity*, or *firmness*, or *steadiness*" (T 1.3.7.7; SBN 629). And further,

> This force and this vivacity [of ideas] are most conspicuous in the memory; and therefore our confidence in the veracity of that faculty is the greatest imaginable, and equals in many respects the assurance of a demonstration. The next degree of these qualities is that deriv'd from the relation of cause and effect; and this too is very great, especially when the conjunction is found by experience to be perfectly constant, and when the object, which is present to us, exactly resembles those, of which we have had experience. But below this degree of evidence there are many others, which have an influence on the passions and imagination, proportion'd to that degree of force and vivacity, which they communicate to the ideas.
>
> (T 1.3.13.19; SBN 153)

Because they concern feeling, the distinctions Hume makes are like aesthetic distinctions.

> 'Tis not solely in poetry and music, we must follow our taste and sentiment, but likewise in philosophy. When I am convinc'd of any principle, 'tis only an idea, which strikes more strongly upon me. When I give the preference to one set of arguments above another, I do nothing but decide from my feeling concerning the superiority of their influence.[14]
>
> (T 1.3.8.12; SBN 103)

Sometimes ideas are imposed on us by principles of reasoning that are "changeable, weak, and irregular." Their influence can be undercut by a due contrast with ideas imposed on us by principles that are "permanent, irresistible, and universal" (T 1.4.4.1; SBN 225). In this way, "we might hope to establish a system or set of opinions, which if not true (for that, perhaps, is too much to be hop'd for) might at least be satisfactory to the human mind, and might stand the test of the most critical examination" (T 1.4.7.14; SBN 272). Thus, Hume is able to distinguish, on the one hand, views that would be stable through time and from place to place, and, on the other hand, views that would vary by time or place. The latter would include superstitions, myths, the fictions of the ancient philosophers. Some of our stable views will be fundamental natural beliefs, such as those in the unitary self and the external world, but there is also room for views in philosophy, science, and mathematics. These latter might even yield views opposing natural ones, as, for instance, the philosopher's hypothesis of "the double existence of perceptions and objects" opposes the natural view that "[o]ur perceptions are our only objects" (T 1.4.2.50–2; SBN 213–15). And so Hume makes room for "refin'd reasoning" and "the most elaborate philosophical researches" in his skeptical approach.[15] Like a critic in the arts who tries to distinguish classic works from passing fancies, Hume tries to distinguish the most stable of the views forced upon us by appearances. And so, in yielding to arguments that feel more stable, Hume engages in an analogue to theorizing, but without any attempt to get at the truth behind appearances. Thus, Hume is a Pyrrhonian skeptic with respect to assent, whose discrimination concerning the force and stability of passive acquiescence in views, gives him a procedure for doing philosophy, science, and mathematics with results akin to those of the Academic mitigated skeptics and their modern heirs. The difference is that Hume makes no connection between the degree to which a view is forced upon him and the degree to which it is likely to be true.

This appeal to stability gives Hume a surrogate, naturalistic account of normativity and reasonableness, one that enables him, for instance, to recommend philosophy over superstition (T 1.4.7.13; SBN 271), and to recommend principles of good induction (T 1.3.15; SBN 173–6).[16] The normative force of the recommendations, such as it is, is ultimately provided by our natural and almost universally shared interest in surviving, in increasing pleasure, and in reducing pain (T 1.4.4.1, 1.4.7.11–12; SBN 225–6, 270–1).[17] Hume's bet is that stable views promote these interests.

Much success in arriving at stability is to be hoped for when our views concern sensory appearances of objects. There are views provoked by other sorts of appearances, but few as stable: "As long as we confine our speculations to the *appearances* of objects to our senses, without entering into disquisitions concerning their real natures and operations, we are safe from all difficulties, and can never be embarrass'd by any question" (T 1.2.5.26 n.12; SBN 638).

Note the appeal to objects as they appear to the senses. Hume's skepticism explains his empiricism. That is, his assent only to views forced on him by appearances explains his reliance on experience. He says that "we can never pretend to know body otherwise than by those external properties, which discover themselves to the senses." When discussing them he contents himself "with knowing perfectly the manner in which objects affect my senses, and their connexions with each other, as far as experience informs me of them" (T 1.2.5.26; SBN 64). This approach is the one he promises in introducing his "science of man." Thus he says that "the only solid foundation" for his theory of human nature "must be laid on experience and observation" (T Intro.7; SBN xvi). This empiricism, which he learned from Boyle, perhaps Newton, and others, Hume takes simply to be an aspect of his Pyrrhonism:

> And tho' we must endeavour to render all our principles as universal as possible, by tracing up our experiments to the utmost, and explaining all effects from the simplest and fewest causes, 'tis still certain we cannot go beyond experience; and any hypothesis, that pretends to discover the ultimate original qualities of human nature, ought at first to be rejected as presumptuous and chimerical.
>
> (T Intro.8; SBN xvii)

As with the study of human nature, so with all sciences. "None of them can go beyond experience, or establish any principles which are not founded on that authority" (T Intro 10; SBN xviii). Hume's Pyrrhonian Empiricism is thus concerned only with objects as they appear to us in experience.[18]

Some incautious uses of the words "real" and "really" might be thought to belie this claim.[19] When characterizing his skepticism, he contrasts the appearances of objects with their unknowable real natures, as one would expect. However, once he takes it to be understood that he is confining his attention to these appearances, he sometimes feels free to reapply the contrast between appearance and reality. In effect he is distinguishing how an object really appears from how it apparently appears—a contrast allowed by his distinction between more stable and less stable views of things. Other times, he contrasts being real with being a nonentity, in other words, being something that appears as opposed to being nothing. Sometimes he uses "in reality" to distinguish a view forced on the mind from the view being criticized. Sometimes he contrasts being real with being only in the mind, though again this contrast is within the world as it appears. Such uses of "real" "imply no dogmatical spirit" anymore than phrases such as "*'tis evident, 'tis certain, 'tis undeniable.*" The propensity to use such expressions is "so natural." As Hume says, they were "extorted from me by the present view of the object" (T 1.4.7.15; SBN 273–4). Despite such expressions, he is still confining his attention to objects as they appear to us in experience.

One of Hume's innovations was to find out about objects as they appear to us, by examination of the ideas we use to represent them. For instance, he comes by this means to the conclusion that a mathematical point must have color or solidity in order to be an entity that can "by its conjunction with others form a real existence" (T 1.2.4.3; SBN 40).[20] In the course of this investigation he asks, "*What is our idea of a simple and indivisible point?*" and comments, "No wonder if my answer appear somewhat new, since the question itself has scarce ever yet been thought of. We are wont to dispute concerning the nature of mathematical points, but seldom concerning the nature of their ideas" (T 1.2.3.14; SBN 38).

Examining our ideas can help in finding out how objects appear to us. When our ideas are obscure, we can settle controversies about them by examining the impressions from which the parts of the ideas are copied (T 1.2.3.1; SBN 33). Hume's famous copy principle—that all simple ideas are copied from impressions—is part of his skeptical view that the ultimate source of all our views is appearances (see T 1.1.1.7; SBN 4).

All the theorizing in his science of man seems to take Hume far beyond his Pyrrhonian roots. Sextus says, "For anyone who holds beliefs on even one subject, or in general prefers one appearance to another in point of convincingness or lack of convincingness, or makes assertions about any unclear matter, thereby has the distinctive character of a Dogmatist."[21] However, it seems to me Sextus is here dogmatically making an assertion about an unclear matter, unless he is read as merely passively acquiescing in this view concerning other people's attempts to actively endorse views or at least come to reasonable conclusions. Concerning his theoretical views, Hume can be read the way Sextus is to be read, and not the way Sextus is reading his opponents. If, in doing the things listed, Hume is passively acquiescing in views forced on him by appearances, then he is not being dogmatic.[22]

As Pyrrhonian, Hume does not even actively endorse the philosophical framework he uses to characterize his skeptical approach. That there are external physical objects (what Hume calls in general "body") is not something Hume finds any justification to believe. It appears, upon review, to be based on a "gross illusion" (T.1.4.2.56; SBN 217). Rather, belief in body is a belief forced upon us. As he says, "'tis in vain to ask, *Whether there be body or not?* That is a point, which we must take for granted in all our reasonings" (T 1.4.2.1; SBN 187). Further, that there is an internal world of perceptions caused by the external world of objects, is a belief philosophers find themselves with when they reflect on experience: "For as the philosophical system is found by experience to take hold of many minds, and in particular of all those, who reflect ever so little on this subject, it must derive all its authority from the vulgar system; since it has no original authority of its own" (T 1.4.2.49; SBN 213).

Philosophers only believe in inner perceptions as a result of simple experiments showing that perceptions depend on our organs of sense. For example, one can double one's visual impressions by gently pressing one eye (1.4.2.45; SBN 210–11). However, the experiments yield such a conclusion only assuming that there are physical objects, some of which are organs of sense.[23] Thus, the belief in inner perceptions has no more justification that the belief in body.[24] Belief in both is, nonetheless, durable for those prone to philosophy (1.4.2.57; SBN 218).

One might focus on passages in which Hume is questioning this philosophical framework, or is thinking along the lines of ordinary non-philosophical people, to conclude that by talk about the external world he means really to be talking about our impressions of it. But when doing philosophy one cannot shake the framework for long (T 1.4.2.41–57; SBN 208–18).[25] When Hume is operating within it, he means really to be talking about the external world as it is experienced via our impressions and as it is represented by our most stable ideas. The framework may seem for a moment to be less justified than the conclusion that there are only perceptions, but the framework is more stable.

We might be misled by Hume's claim that his philosophy "pretends only to explain the nature and causes of our perceptions, or impressions and ideas" (T 1.2.5.26; SBN 64). We are more easily misled if we confuse objects as they appear with the impressions they occasion. Certainly Hume devotes attention and care to characterizing our perceptions. But the context of the above quotation is a paragraph in which he is contrasting what he is *not* trying to do—"penetrate into the nature of bodies"—with what he is trying to do—to "know body ... by those external properties, which discover themselves to the senses." Part of the task in examining perceptions is to use what he learns in order to characterize the objects of those perceptions, that is, objects as they appear to us. In other words, as a Pyrrhonian Empiricist, he is concerned to characterize the external world in those appearances that force views upon us.[26]

Metaphysician and skeptic

I have emphasized Hume's Pyrrhonian Empiricism as a way of resolving the traditional tension in interpreting Hume's approach. There could well be more to Hume's Pyrrhonism than acquiescing in views forced on one by sensory appearance. He could additionally acquiesce in views held by tradition. As Sextus says,

> Thus, attending to what is apparent, we live in accordance with everyday observances, without holding opinions—for we are not able to be utterly inactive. These everyday observances seem to be fourfold, and to consist in guidance by nature, necessitation by feelings, handing down of laws and customs, and teaching of kinds of expertise.[27]

Concerning the third he says, "By the handing down of customs and laws, we accept, from an everyday point of view, that piety is good and impiety bad."[28] I am assuming that traditional beliefs held by custom can likewise be handed down.

How would traditional views have their effect on one? Perhaps Hume would say that, as in the case of education or the oft-repeated lie, repetition builds assent (T 1.3.9.19; SBN 117). The repetition of traditional beliefs impresses them forcefully on the mind. This is not to say, of course, that traditional beliefs cannot be given up. Just as contrary appearances can be weighed and one triumph, so tradition can be weighed against appearance or contrary tradition. The point is, though, that the skeptic who passively acquiesces in views will tend to hold traditional beliefs. Someone imbued with philosophy will be moved by traditional metaphysical beliefs that are not contrary to experience. Hume can be interpreted as an example of this phenomenon. He uses some metaphysical principles which are well established even if not uncontroversial, and which, he seems to assume, are controversial only because not fully understood. For instance he appeals to "the common sentiment of metaphysicians" at T 1.2.2.3; SBN 30, and "an established maxim in metaphysics" at T 1.2.2.8; SBN 32.

It might seem that respecting tradition in the way I suggest, reopens the door to the rejected divinity and school metaphysics. If tradition gives a view some force, then perhaps some of these rejected views backed by tradition will outweigh ones Hume recommends. In principle, yes, perhaps they might. However, Hume is confident that exposing sophistries and illusions, even ones backed by tradition, will undermine all but the hardiest beliefs. For instance, Hume argues that the traditional tenet that God is incomprehensible amounts to skepticism or atheism, or the traditional tenet that God is a simple, thinking mind is inconsistent, since thinking minds are necessarily complexes of perceptions (*Dialogues*, Part 4). One tenet, however, namely that God is a mind, will survive counterarguments. In this case, he thinks it is because of the naturalness of the argument from design, though, and not simply because of being traditional.

Not all of Hume's metaphysical assumptions can be said to be simply traditional. Where a tradition seems wrong (see T 1.2.5.22, 1.4.5.35, 1.4.6.4; SBN 62, 250–1, 252–3), where traditions conflict, or where there is no established tradition, a choice has to be made. Here, a skeptic can appeal to his metaphysical taste. Hume can agree with Peirce that a-priori theorizing with nonanalytic propositions is a matter of taste.[29] He can develop his metaphysical taste by carefully weighing opposing positions and considering their consequences for empirical and mathematical science. He can then gamble that he has developed that taste to the extent that the metaphysical propositions that strike him most forcefully will be the ones that will prove most stable.

> 'Tis not solely in poetry and music, we must follow our taste and sentiment, but likewise in philosophy. When I am convinc'd of any principle, 'tis only an idea, which strikes more strongly upon me. When I give the preference to one set of arguments above another, I do nothing but decide from my feeling concerning the superiority of their influence.
> (T 1.3.8.12; SBN 103)

In "Of the Standard of Taste," Hume explains that "the principles of taste" are "universal, and nearly, if not entirely the same in all men" (OST 241). The few who have the requisite sense, delicacy, lack of prejudice, and practice are the ones who can tell which works have the qualities that, given these "general principles of approbation or blame," will give rise to "durable admiration" that will survive "all the caprices of mode and fashion, all the mistakes of ignorance and envy," and "[a]ll the changes of climate, government, religion, and language" (OST, 233).

A skeptic such as Hume can suspect that there are widespread principles of metaphysical taste, of course brought to full fruition only in a few philosophers. These principles might allow for a certain amount of variation, as in art (OST, 243), but would still be the basis for some views becoming metaphysical classics that stand the test of time.

In relying on passive acquiescence to tradition or to developed taste, in striving for stability, and in seeing no reason to think that stability indicates truth, Hume can maintain his skepticism while engaging in metaphysics.

2 Moments and durations

Hume's discussion of numerical identity presupposes his discussion of time, with which I begin. On Hume's view of time, the things in time are either temporal simples or temporal complexes composed of temporal simples. A temporal simple has no parts occupying distinct moments, whereas a temporal complex does.[1] Something is a temporal simple if and only if it occupies just a single moment. If it occupies just a single moment then it lacks duration, so something has duration only if it is a temporal complex.

On Hume's general view of parts and whole, a complex is not really a single thing; it is really just its many distinct parts.[2] Something without parts, on the other hand, is a single thing. Thus, a temporal simple is a single thing. Something with duration, in contrast, is really many distinct things. It is this contrast in his account of time that he appeals to in giving his account of the idea of identity.

To show that Hume held these views, I will first discuss his account of the idea of time as an abstract idea derived from ideas of successions. The discussion of abstraction will show that, just as successions are wholes with their members as parts, so durations are wholes with moments as parts; that all and only successions have duration (that is, occupy durations); and that individual members of successions occupy individual moments. Then I will discuss two of his proofs that there are some parts of time without parts. These two proofs are among his various arguments that finite portions of space and time are not infinitely divisible. The discussion of infinite divisibility will show that durations and successions, being wholes, are really many things, and that moments lack parts and so are single things.

Time and abstraction

For Hume, time is an abstraction from the successions we experience (T 1.2.3.6–11, 1.2.4.2; SBN 34–7, 39–40). That is to say, the idea of time is an abstract idea of any succession qua many things in succession. Likewise I will propose that the idea of a moment is the idea of a member of a succession qua member.

An abstract idea, for Hume, as for Locke, is an idea of an object with a characteristic (or some characteristics) shared by a number of particular things that differ in other respects. The resemblance enables the abstract idea to represent each of the particulars all at once. Paradigmatically, in virtue of sharing the relevant characteristic(s), the particulars are of the same sort.[3] Having general representation is the only way the abstract idea is general.[4] And, paradigmatically, it is impossible for anything real to have just the relevant characteristic(s) and no others, where the others are paradigmatically "particular degree[s]" (T 1.1.7.2; SBN 17) of the abstract characteristic(s).[5]

These assumptions are shared by Hume and Locke. Locke provided the received account of abstract ideas—deriving ultimately from Boethius—which Hume, following Berkeley, challenges.[6] The last of the shared assumptions—that it is impossible for anything real to have just the relevant characteristic(s)—is used to defeat Locke's account. Locke held that the only way an abstract idea could represent many differing particulars is by being of an object that has only the relevant characteristic(s) and none of the differing particular degrees. That is, the abstract idea has to be of an object from which irrelevant characteristics have been separated. In contrast, to this question "whether abstraction implies a separation" (T 1.1.7.3; SBN 18), Berkeley and Hume answer, No. They contend that to have an idea of such an object is to conceive of it. What is conceivable is possible in reality. So such an object is possible in reality. But all agree it isn't possible. (That's the last of the shared assumptions.) So there is no such idea.[7]

After this negative result, the constructive task is to devise an account of abstract ideas that satisfies *all* the criteria shared with Locke, even the last which defeated Locke's account. Thus, the task is to give an account in which abstraction does *not* imply "a separation." Berkeley began such an account, drawing on the traditional notion of partial consideration, and Hume refined it. Hume's account is this: Thinking of a sort of thing requires a word, an idea, and a habit. The word is the name of the sort—a "general term"—such as "a man," "a line," "a triangle," "government," "church," "negotiation," "conquest," etc.[8] Using the word causes an idea of one of the particular members of the sort to be present to mind. Additionally, using the word engages a habit of bringing to mind, as needed, any idea the user has of any of the other particular members. Paradigmatically, the need is to prevent rash claims about that sort of thing based just on the idea present to mind. For example, if one is tempted to say, based on an idea of an equilateral triangle present to mind, "that the three angles of a triangle are equal to each other," one is immediately prevented by the ideas of some counterexamples (T 1.1.7.8; SBN 21). Or, for example, if one's prejudices about Irishmen or Frenchmen are not completely unreasonable, one will think of counterexamples when expressing them (see T 1.3.13.7; SBN 146–7).

It is the perceived resemblance of the members of the sort that enables the mind, when thinking of one member, to tend to think of the rest (or at least those of which one has ideas). And it is this resemblance that enables the mind to associate the same word[9] with each of them. The members are considered in the light of this resemblance, as Hume would say. In other words, when the idea of each is before the mind, one pays attention only to the respect of resemblance and fails to attend to the differing degrees of it and to other ways in which the members of the same sort differ. This is the sort of selective attention that makes possible a distinction of reason between the respect in which a thing resembles things of one sort, and the respect in which it resembles things of another. Selective attention needs to be distinguished from the "separation by the mind" championed by Locke and attacked by Berkeley and Hume. In the latter, the separation, the idea present to the mind is of something that *lacks* the degrees. In the former the idea is of something that *has* the degrees, but they are not attended to.

Thus the abstract idea of a human, is the idea of a particular human considered in the light of his mutual resemblance to all humans (or at least all those of whom one has ideas) and used to represent every human. This is what I mean by the phrase "the idea of any human qua human."

With such an account, the criteria for an abstract idea are met. Thus, Hume can legitimately speak of abstract ideas or ideas of abstract objects. For example, the abstract idea of a man is a particular idea of a particular man serving as an idea of any man qua man. Hume applies this account in his discussion of the idea of time.

In that discussion, the relevant particular ideas are ideas of particular things in succession, for example, "five notes play'd on a flute" (T 1.2.3.10; SBN 36). Such a succession resembles other successions in at least one respect: successiveness.[10] When this has been noticed and the sort named, one has an idea of any succession qua successive. The obvious general term here is "a succession."[11] However, we have come to use another also: "time." The idea of time, then, is the idea of any succession qua successive. He says of our "conception of time" that it "can plainly be nothing but different ideas, or impressions, or objects dispos'd in a certain manner, that is, succeeding each other" (T 1.2.3.10; SBN 37). When we say something has occurred in time, Hume would have us understand that something has occurred in a succession. When we say time has passed, Hume would have us understand that a succession has passed.

Hume seems to use "time" and "duration" interchangeably within T 1.2.3.6–11, 17; SBN 34–7, 39. This makes sense when speaking of *a* duration. Thus "time," "a succession," and "a duration," when used generally, are interchangeable for Hume. However "duration" can be used also to convey more the manner than the sort, more successiveness than a succession. Hume tends not to use it in this sense but commentators often read it this way, perhaps influenced by Kemp Smith.[12]

What Hume says about time is parallel to what he says about space. The idea of space is the idea of any extension qua extended. As he says "the idea of extension is nothing but a copy of these colour'd points, and of the manner of their appearance." That they are of some particular color is, of course, not part of the manner represented because we are considering them in the light of their resemblance to other arrays of points both seen and merely felt (T 1.2.3.4–5; SBN 34). Note that, on Hume's view, time and space are not dimensions or containers of events. They are abstractions from perceived arrays of individuals.

A tempting objection to all this is that Hume's account of the abstract idea of time is circular. It would seem that Hume would need us to have the idea of successiveness in order to discern and attend to the successiveness in successions of perceptions.[13] However, this charge of circularity is misguided. A certain point of resemblance might intrude on our attention if we are naturally susceptible to the intrusion.[14] And this natural susceptibility would not have to be the same thing as already having an idea of the point of resemblance. That is, current particular successions might naturally tend to bring to mind past ones. And acquiring this new habit of association might naturally cause us to assign a word to evoke it in the future. After all, we are just causal systems according to Hume's science of man (T Intro.8; SBN xvii). The result is an abstract idea—viz., a particular idea (here, of a succession), plus a word, plus a habit (T 1.1.7.10; SBN 22). As long as successiveness naturally intrudes on our attention and we naturally find it convenient to coin a word for it to remind us of things that resemble with respect to successiveness, Hume's account is noncircular. We can arrive at the abstract idea of a succession in general without already having had it.

Rosenberg accuses Hume of another sort of circularity. He says that Hume uses the concept of succession in his analysis of the concept of time, but since succession is a temporal concept, the analysis is circular.[15] This charge would succeed if Hume were giving an analysis of the concept of time, but Hume gives little reason to suppose that this is what he is doing. Rather, he is explaining how we arrive at the idea of time. He says that when we have successions of perceptions, the fact that they resemble with respect to successiveness causes us to form an idea of a succession in general. No circularity here.

So time, in other words duration, is a succession in general with special attention to the manner in which successions are arrayed—successiveness.

A direct consequence of Hume's account is that time or duration has parts. Successions have parts, namely their members, so a succession in general has parts. In this respect, a duration is just like an extension, an analogy Hume relies on when discussing infinite divisibility (see T 1.2.3.5, 1.2.3.16; SBN 34, 39). The parts of time are moments (T 1.2.2.4; SBN 31).

Hume's explicit argument that time has parts gives short shrift to a worry raised by St. Augustine. Hume says, "'Tis evident, that time or duration consists of different parts: For otherwise we cou'd not conceive a longer or shorter duration" (T 1.2.3.8; SBN 35–6). Likely he has in mind Augustine's worry that, since past and future do not exist, time consists only of the fleeting present moment, and so there cannot be such a thing as a long time.[16] Hume rejects the consequent and so concludes that there is more to time than the present moment and so time has distinct parts. Successive parts of time exist, just not all presently. Apparently Hume rejects Augustine's assumption that what does not exist presently does not exist in any way, and so cannot be part of time.

Another direct consequence of Hume's account is that *all* successions have duration. Since the idea of duration is the idea of a succession qua successive, it applies to any succession. It is true that Hume says that the idea applies only to what it can be derived from (T 1.2.3.11; SBN 37), and that it can be derived only from a *perceivable* succession (T 1.2.3.7; SBN 35). However, it does no harm to take the idea to apply also to imperceivable successions. It is likely Hume intended us to do that, since he also says "that time is nothing but the manner, in which some real objects exist" without insisting that the manner be perceivable (T 1.2.5.28; SBN 64).

Yet another direct consequence of Hume's account is that *only* successions have duration. It is clear that objects which exist only for a brief moment do not have duration, but the difficult case is objects that are "stedfast and unchangeable" (T 1.2.3.11; SBN 37). These are cases "without any change or succession" (T 1.2.5.29; SBN 65). In such a case there is just a single object, not distinct objects successively. So the idea of a steadfast object cannot serve as an idea of duration. So a steadfast object cannot convey the idea of duration. So a steadfast object lacks duration: "For it inevitably follows from thence, that since the idea of duration cannot be deriv'd from such an object, it can never in any propriety or exactness be apply'd to it, nor can any thing unchangeable be ever said to have duration" (T 1.2.3.11; SBN 37). Only by a fiction can one come to think of a steadfast object as having duration.[17] So it is Hume's view that all and only successions have duration.

A further consequence of Hume's account, one he does not explicitly draw, concerns what a moment of time is. To address this question I need again to draw attention to the way my interpretation differs from that of Kemp Smith and others.[18] In these accounts the idea of time is the idea of a manner, viz., the manner in which successions are arrayed. Hume certainly talks this way at T 1.2.4.2; SBN 39–40. Given Hume's account of abstraction, however, it is better to think of the idea of time as the idea of a succession in general. Yes, the mind "takes notice of the *manner*," and there is no particular succession of objects which the mind must think of in order to think of the manner. Nonetheless "[t]he ideas of some objects it certainly must have" (T 1.2.3.10; SBN 37). For Hume there is no distinction between the idea of a manner and the general idea of objects arrayed in that manner. For us who make the distinction, however, Hume's idea of time is more perspicuously thought of as the latter. There is an added advantage to my view. Thinking of time as a manner makes it hard to see what parts of time could be. How would a manner have parts? But surely it is not hard to conceive how a succession in general has parts.[19]

The idea of time is derived from successions. So it stands to reason that the idea of a moment is the idea of a part of a succession. It would be more idiomatic to say a "member" of a succession rather than a "part." So, in accordance with the account of the abstract idea of time just given, I propose that the idea of a moment is the idea of any member of a succession qua member of a succession. Thus, strictly speaking, the general term "a moment" and the general term "a member of a succession" can be used interchangeably when thinking abstractly.

Given this account of being a moment in general, being a particular moment would be a matter of being this member of this succession qua this member of this succession. I don't mean being this member qua itself; what I mean is being this member qua later than these members, earlier than those members, and coexistent with these members of other successions. The idea of a particular moment would still be an abstract idea, since it would be possible to use an idea of a distinct particular object standing in all the same temporal relations to represent the same particular moment.

Given this account of particular moments, a member of a succession occupies the particular moment that can be abstracted from it. After all, to think of time requires thinking of a succession with an eye to the manner in which its members are arranged, that is, with an eye to its successiveness. "The ideas of some objects [the mind] certainly must have" (T 1.2.3.10; SBN 37). The moments are said by Hume to be "fill'd" with these objects (T 1.2.3.17; SBN 39). In other words, the object occupies the particular moment that is abstracted from it.[20]

These, then, are what moments are, given that time is an abstraction and moments are its parts. The next step is to argue that moments are not composed of briefer moments and so are single things, whereas durations and successions are really many things.

Time and infinite divisibility

Nowadays, following Cantor, we think that for a finite spatial or temporal interval to be infinitely divisible, can be for it to have an infinity of indivisible parts (in some sense of "part"). Although his main argument against infinite divisibility also rules it out in the above sense, for Hume, being infinitely divisible means every part being divisible into parts.[21] It is this conception that he contends against in the *Treatise* and continues to contend against in the first *Enquiry.*

> A real quantity, infinitely less than any finite quantity, containing quantities infinitely less than itself, and so on, *in infinitum;* this is an edifice so bold and prodigious, that it is too weighty for any pretended demonstration to support, because it shocks the clearest and most natural principles of human reason.
>
> (EHU 12.18; SBN 156)

Rather, "'Tis certain then, that time, as it exists, must be compos'd of indivisible moments" (T 1.2.2.4; SBN 31). I will focus on two arguments that best show the connection Hume takes there to be between unity and simplicity—the Malezieu argument at T 1.2.2.3; SBN 30–1 and the Additional argument at T 1.2.2.4; SBN 31.[22] We need to take them seriously despite widespread conviction that they are laughably flawed. The main attacks on these arguments come in Antony Flew's highly influential paper, so along the way I will show that his attacks are misconceived. Likewise, considerations raised by Fogelin and Laird are, for these two arguments, beside the point.[23] Ironically, some of the criticisms show neglect of Cantor's conception of a line as an actual infinity. It is strange that Cantor could be used to defend Hume.

My focus here is time, so I will talk about space only to help clarify the arguments about time. Hume argues that moments of time are not composed, even potentially, of briefer moments.

Both Hume's arguments here depend on what I will call the Divisibility Assumption. So I will begin by defending it, before turning to the details of each argument.

Hume takes to be obvious that *anything divisible, actually (not just potentially) has parts.* This assumption is almost explicitly stated at T 1.2.2.2; SBN 29–30 where he says, "Everything capable of being infinitely divided contains an infinite number of parts." I'll call this quoted claim the Infinite Divisibility Assumption.[24] It is hard to reject these assumptions for moments specifically, because it is hard to know what dividing a moment could be other than discerning the parts that it actually has. Still, the Divisibility Assumption has been contradicted by a long tradition going back to Aristotle who concluded that something divisible need not have parts that exist actually, but need only have parts that exist potentially.[25] So a rejoinder is needed, not only to support the assumption but also to explain why Hume takes it to be obvious.[26]

An argument of Bayle's can provide inspiration. Bayle argues against extended but indivisible "Epicurean atoms" by observing, "I can deny concerning the right side what I affirm about the left side. These two sides are not in the same place."[27] Following this lead, the following rejoinder can be given to the Aristotelians on Hume's behalf: Some—but not all—of a whole would become its left half, were the whole divided in half. The rest would become the right half. The former is actually on the left and the latter is actually on the right. They actually differ. Since something cannot differ from itself, they are actually numerically distinct. Since these lesser amounts of the whole are actually distinct from each other, then they are actually parts of the whole. So anything divisible actually has the parts it is divisible into.[28]

Flew, without explicitly distinguishing them, has mocked Hume for holding both the Divisibility Assumption and the Infinite Divisibility Assumption. My concern is just the first one. But the first, with the plausible assumption that something has all the parts that would result from any division, entails the second. So I need to answer arguments against the second also. I will show that Flew's two arguments are based first on confusion and second on misunderstanding.[29]

First, Flew says that obviously the divisible is not thereby divided. A cake merely divisible into slices is not thereby divided into slices. This truth would seem to constitute an attack on Hume's principle only if one confused having numerically distinct parts with being divided into parts. But this is a confusion to avoid. Undivided parts can be numerically distinct. In fact, as I have argued, they have to be. This is the point of Hume's Divisibility Assumption. As Bayle puts it, "The continuity of parts does not prevent their actual distinction."[30] Flew's truism is irrelevant to the claim that the divisible has parts.

Second, Flew contends that to be infinitely divisible does not entail the possibility of a division into an infinity of parts. Flew seems to think that Hume needed a process of division to be both endless yet have an end in order for his Infinite Divisibility Assumption to hold. But as I have shown, the Infinite Divisibility Assumption just entails that being infinitely divisible entails having an infinity of distinct parts. They can exist independent of any process of division. Flew's point, again, is irrelevant.

Additionally when Flew exclaims that infinity is not a number, he seems to be neglecting Cantor's teachings about transfinite numbers. Further, if Flew thinks, as he seems to, that there is no good sense in which something infinitely divisible has an infinite number of parts, then he is wrong. On the Cantorean conception of a line as a set of points with the order type of the continuum, there is a perfectly good sense of "part" in which the points are parts. They are elements. Alternatively if you mean by "point" a unit point-set, then points are subsets. These would also be parts in a perhaps even better sense of "part."[31] Furthermore there is a transfinite number ($\aleph_1$, if we assume the continuum hypothesis) that gives the cardinality of this set.[32]

Even if we do not like points, there is also another way in which an infinitely divisible interval has an infinity of parts. Let the first part be half the interval, the next part be half the remainder, the next part be half the new remainder, and so on. There will be a countable infinity of these parts. The cardinality of this set of parts will be $\aleph_0$.[33] On the Cantorean conception of a line, these parts will be subsets, which again is a perfectly good sense of "part." So whether we consider points or these proportional parts, Hume's Infinite Divisibility assumption is upheld.

I am not saying that these objections to Flew would be endorsed by Hume. He thought that we could know by examining our minutest ideas that the kind of order needed for a continuum is impossible (T 1.2.2.2; SBN 29–30). But that is a separate issue independent of the truth of the Divisibility and Infinite Divisibility Assumptions. I am just saying that we ourselves should not reject these Humean assumptions on the basis of Flew's arguments.

Fogelin also chastises Hume for holding the Infinite Divisibility Assumption.[34] He claims that Hume, in thinking the infinitely divisible has parts, fails to distinguish two senses of "part" that we moderns distinguish. Fogelin doesn't say what those senses are. Perhaps he has in mind a distinction between part, properly so-called, and limit.[35] Presumably the idea is that parts can be arrived at by division, but limits cannot be. There are scraps of evidence that this is what Fogelin had in mind in his later article. He says, "the believers in infinite divisibility deny that a line is composed of fundamental parts—a line is divisible all the way down without a stopping place."[36] And, later in the article, Fogelin mentions, concerning Berkeley, that some of his arguments "depend upon misunderstandings of the notion of a limit."[37] In any event, the way of thinking of things in Fogelin's criticism of Hume derives, very likely, from Aristotle's answer to Zeno's paradox of extension.[38] However, this way of thinking neglects the point I made above: Whether parts or limits can be arrived at by a division is beside the point. All that matters is numerical distinctness. Fogelin's assumption that lacking fundamental parts is equivalent to being divisible all the way down, as in the first quotation, is a mistake. Something with fundamental parts can nonetheless be infinitely divisible, as we have seen. Or if Fogelin's assumption is not a mistake, that would only be because he restricted the sense of "part" to something that can in principle be arrived at by division. But this restriction would be unfair given the Cantorean conception of a line. On that conception there is a perfectly good sense of "part" in which a limit is a part.

Thus the Divisibility Assumption is not obviously wrong and in fact is likely right.

Now for Hume's arguments that connect unity and simplicity. According to Hume, moments are indivisible. The Malezieu argument at T 1.2.2.3; SBN 30–1 and the Additional argument at T 1.2.2.4; SBN 31 support this conclusion.

Although cast as an argument about extension, the Malezieu argument concerns the metaphysics of parts and wholes in general.[39] This concern is entirely appropriate here given Hume's view that our conception of time is the abstract idea of any whole with successive parts, as such. My primary concern is to show the strength of Hume's metaphysical arguments. Along the way I will also show that standard objections that Hume is simply making mathematical mistakes are not well founded.

The Malezieu argument is "a strong and beautiful" argument according to Hume (T 1.2.2.3; SBN 30). Hume's first assumption is the Divisibility Assumption.

His next assumption is that *anything with parts is many things, not a single thing*. Call this the Plurality Assumption.[40] This assumption is revealed in his claim that "extension is always a number, according to the common sentiment of metaphysicians" (T 1.2.2.3; SBN 30).[41] This claim partly relies on something stated explicitly elsewhere: "Whatever is extended consists of parts" (T 1.4.5.7; SBN 234). The link between this statement and the first quoted claim is the Plurality Assumption that anything with parts is a number of things.

It is somewhat hard to understand what the sentence "Anything with parts is many things, not a single thing" means. Forget metaphysical theories for a moment and consider this question: Is a crowd one thing or many things? A natural answer is that it is many things—i.e., many people—not a single thing the way a person is a single thing. Why? Because if there are no people, then there is no crowd. If a metaphysician tries to capture this natural answer he may well say that a crowd is not a single thing that depends for its existence on standing in a composition relation to many people. Rather, a crowd just is many people close together.[42]

Thus, a grammatically singular expression such as "a crowd" or "a succession" or "a plurality" can be used to refer collectively to many things in some relation. Why would we have grammatically singular expressions to refer to what are many? Presumably because it is convenient to consider them as one thing when the distinctions between the members are unimportant.[43] It is convenient but, according to Hume, strictly false. Thus, a grammatically singular expression falsely conveys, as a matter of convenience, the singularity of what it refers to. To tell the literal truth as best we can, given the materials we have to work with, we start out with the convenient way of referring, then say what is true about what is referred to. So it makes sense to say that something is many things in relation. That is what is being said about something with parts: it is just the many parts connected somehow.

Hume's third assumption is that *only single things really exist*. Call this the Existence Assumption. "'Tis evident, that existence in itself belongs only to unity" (T 1.2.2.3; SBN 30). Something that is many things does not literally exist (singular). It is not an it; it is a they. "It exists" said of it is literally false. "They exist" is what is literally true. So, for example, to say "A crowd exists" (or perhaps more idiomatically "There is a crowd") is not literally true. We take it to be true because we take it to convey something that is literally true, namely, "Many people each exist and are close together" (or more idiomatically, "There are many people close together").

The basis for this third assumption is that the existence of something that is a number of things depends entirely on the existence of (and relations between, I think it should be said) the things in that number. "Twenty men may be said to exist; but 'tis only because one, two, three, four, *&c.* are existent; and if you deny the existence of the latter, that of the former falls of course" (T 1.2.2.3; SBN 30).[44] Were it possible for "a number of things exists" to be literally true, then the number of things could exist by itself, without the units it consists of. As Hume says, the definition of "substance" ("*something which may exist by itself*") "agrees to every thing, that can possibly be conceiv'd" (T 1.4.5.5; SBN 233). However, it is impossible that a number of things exist without the things in that number.

The three assumptions I have highlighted form part of the basis for Hume's Malezieu argument:

> 'Tis therefore utterly absurd to suppose any number to exist, and yet deny the existence of unites; and as extension is always a number, according to the common sentiment of metaphysicians, and never resolves itself into any unite or indivisible quantity, it follows, that extension can never at all exist.
>
> (T 1.2.2.3; SBN 30)

This argument should be interpreted as follows:

1. Anything divisible actually has parts.
2. Anything with parts is many things, not a single thing.
3. Only single things really exist.

So, anything divisible does not really exist.

This result and the considerations behind its premises yield the fourth assumption—*that a many can be said with some truth to exist only if the individuals in it exist*. The claim that the many exists would be literally false but would have some truth. Thus, using Hume's example, a crowd of twenty could be said to exist only if this person exists, and that one exists, etc. Since one can say anything, I assume Hume is telling us how to say *with some truth* that the crowd exists. The fifth assumption is the one mentioned at the outset—*that infinite divisibility in the relevant sense is that every part of something has parts, and so on*. Thus the argument continues:

4. Anything that is many things can be said with some truth to exist only if those many things each exist.
5. Anything infinitely divisible is such that all its parts have parts.

So, none of its parts really exist.
So, nothing infinitely divisible can be said with any truth to exist.

Flew urges us to think that Hume tries to go simply from saying that a collection consists of its members to saying that no collection is infinitely divisible.[45] As I have shown, the steps are more complicated and plausible than Flew gives Hume credit for. In essence, like Leibniz, Hume is arguing that anything extended is a collection and that no collection can be composed only of collections. There must be units that are not collections.[46]

Hume made this argument explicitly to argue that extension is not such that every part is divisible into parts. But then he says, "All this reasoning takes place with regard to time" (T 1.2.2.4; SBN 31). That is, Hume's prior arguments, including the one just interpreted, apply *mutatis mutandis* to time, "along with an additional argument, which it may be proper to take notice of."

Hume has just argued that *anything divisible does not really exist*. Now he will go on to argue that *anything temporally divisible does not really exist in the present*.

Flew suggests that the Additional argument doesn't follow and that Hume thought it did only by stubbornly importing the conclusion into the premises.[47] I think, however, that we can read Hume with less disdain.[48]

Hume's argument, like its template in Bayle, takes as obvious that since time is successive, if moments are divisible into moments then some moments coexist.[49] This, at first, seems more perplexing than obvious. The key is to realize that both philosophers take the successiveness of time to entail that moments are present only one after the other. So, were the present moment to have successive moments as parts, the parts would, *per impossible*, be present all together. As Hume puts it, "there wou'd be an infinite number of co-existent moments, or parts of time; which I believe will be allowed to be an arrant contradiction" (T 1.2.2.4; SBN 31). I will explain the argument, then connect my explanation to the way Hume actually puts it.

In making this argument, Hume continues to hold premises (1) and (2) from the previous argument, that anything divisible has parts and that anything with parts is many things. However for purposes of *reductio ad absurdum* he grants his opponents that there really exist moments divisible into smaller moments.[50] In doing so he supposes that premise (3) in the previous argument—that only single things exist—is false. He instead supposes that each moment is "not perfectly single and indivisible" (T 1.2.2.4; SBN 31). Thus each moment has moments as parts, which have moments as parts, etc. Two more assumptions are needed for the argument. First, *only one of successive moments can be present*. This follows from the common-sense view that moments must be present one after the other, which captures an important feature of time as it appears.[51] Second, *all parts of the present moment are present*. After all, if parts of the present moment existed in either the past or in the future, they would be past or future moments, not parts of the present moment. The argument then goes as follows:

> The present moment is divisible into successive moments. (Supposition)
> So, the present moment has an infinite number of successive moments as parts.
>
> 6. Only one of successive moments can be present.
> 7. All parts of the present moment are present.
>
> So all parts of the present moment are present yet only one can be present.
> Since that is a contradiction, the supposition is false.
> So the present moment is not divisible into successive moments.

Hume next assumes that *if not all moments are divisible into successive moments, then none are*. He took the attempt to mix indivisibles and infinitely divisibles to be a subterfuge by advocates of infinite divisibility to avoid admitting clear defeat (T 1.2.4.15; SBN 44). Alternatively he could have argued that if the present moment is not divisible then none are, because (i) the above argument holds for any moment when it is present, (ii) if a moment is not divisible when it is present then it is not divisible, and (iii) all moments are present at some time.

8. If not all moments are divisible into successive moments, then none are.

So, no moment is divisible into successive moments.

The structure of the argument is obscured by the fact that Hume does not explicitly talk about being present, but puts the argument in terms of the existence and coexistence of successive moments. However the argument turns, as Bayle makes clear, on the assumption that moments exist alone: "Every one [i.e., every part of time] has to exist alone. Every one must begin to exist, when the other ceases to do so. Every one must cease to exist, before the following one begins to be."[52]

Since Hume thinks time has distinct parts (otherwise there cannot be longer and shorter times) he thinks some other parts of time exist in the past and perhaps the future. So, in sharing Bayle's assumption, Hume must mean that moments exist alone *in the present*. "[T]ime or succession, tho' it consists likewise of parts, never presents to us more than one at once; nor is it possible for any two of them ever to be co-existent" (T 2.3.7.5; SBN 429). Talk of moments existing is thus talk of their being present. Talk of some moments *co*existing is talk of them all being present. That Hume bases his argument on what is true of a moment when it becomes present is shown by the fact that he talks of "each moment, as it succeeds another" (T 1.2.2.4; SBN 31).[53]

Thus moments lack parts. And so Hume concludes, "'Tis certain then, that time, as it exists, must be compos'd of indivisible moments" (T 1.2.2.4; SBN 31).

It follows that moments are single things, given that Hume uses "unit" and "indivisible thing" interchangeably, as in the phrase "any unite [*sic*] or indivisible quantity" (T 1.2.2.3; SBN 30). Also he says, "But the unity, which can exist alone, and whose existence is necessary to that of all number, is of another kind, and must be perfectly indivisible, and incapable of being resolv'd into any lesser unity" (T 1.2.2.3; SBN 31). In contrast, as seen in the Malezieu argument, anything divisible is really many things. So durations and successions are really many things. The things in time are either temporal simples or temporal complexes. Only the former are single things; only the latter have duration.

3 Steadfast objects

On Hume's view of duration as an abstraction, steadfast objects do not endure. They are temporal simples. Yet he is clear that they coexist with successions, and so do not just briefly exist. But if they exist more than briefly, how can they lack duration? I show that Hume held the strange view that not all temporal simples are uniformly brief and that some temporal simples coexist with successions. Additionally I give the Humean explanation of the view's strangeness. We naturally, but falsely, come to believe that even steadfast objects endure.

Understanding that steadfast objects do not endure but that we hold the unshakeable fiction that they do, is crucial to understanding Hume's account of identity. It is crucial to seeing how the idea of identity is an attempt to reconcile being many with being one.[1]

Being "stedfast and unchangeable" is in contrast to being "a succession of changeable objects" (T 1.2.3.11; SBN 37). A steadfast object, for Hume, is something that is "fast in place, firm; fixed," as Johnson puts it in his dictionary.[2] It does not change; that is, there is no "succession of one thing in the place of another," in Johnson's definition of the relevant sense of "Change." The steadfast object is not quickly replaced, nor is it itself a succession. If it were a succession it would have duration, which according to Hume steadfast objects lack (T 1.2.3.11; SBN 37). Yet while the steadfast object remains unreplaced, other changes occur elsewhere. Not everything is steadfast while it is. Though not being a succession and so lacking duration, it coexists with successions having duration.[3]

The opposition to succession is what is central to Hume's use of "stedfast". There is also the connotation of motionlessness. However, strictly speaking, a steadfast object in Hume's usage can move. For example Hume says the parts of a mass of matter "can continue uninterruptedly and invariably the same" whatever motion they engage in (T 1.4.6.8; SBN 255).[4] Motion certainly gives the idea of duration, but the relevant succession is the succession of places occupied by the object. In such cases "every moment is distinguish'd by a different position" of the object (T 1.2.5.29; SBN 65). This is the way that a case of wheeling a burning coal about is a case of "real succession in the objects" (T 1.2.3.7; SBN 35). There is no succession in the object itself.

Note that having temporal parts entails being a number of things in succession, for Hume. So, not being a succession entails not having temporal parts. When reasoning about space, he makes the analogous premise for this argument explicit. He claims that anything with spatial parts is a number of coexistent things as opposed to a single thing (as opposed to "an unite") (T 1.2.2.3; SBN 30). He then says, "All this reasoning takes place with regard to time" (T 1.2.2.4; SBN 31). The difference is that temporal parts are successive, not coexistent (T 1.2.3.8; SBN 36). So anything with temporal parts is a number of successive things as opposed to a single thing. It is a succession. So a single thing remaining unreplaced lacks temporal parts, because it is not a succession. Thus, Hume thinks that a single thing lacking temporal parts can coexist with a succession of things.

Not only does such a single thing lack actual temporal parts, it lacks potential temporal parts. Anything divisible has parts according to Hume (see T 1.2.2.2; SBN 29). So something lacking parts is indivisible.

Hume's view is even stranger given his view that moments of time are abstractions from single things in time. Each moment is an abstraction from the temporally simple object occupying it. So the structure of temporal relations between single things in time is exactly the structure of temporal relations between moments. Given this and the foregoing, some single, indivisible moments coexist with some successions of single moments. Let me reinforce this conclusion with another argument: Hume thinks that time consists of indivisible moments (T 1.2.2.4; SBN 31). Anything in time exists at one moment at least. Something has duration if and only if it exists at distinct successive moments. So something in time that lacks duration exists at a single indivisible moment. Yet there is something that lacks duration, namely a steadfast object, which coexists with something that has it, namely, some succession. If things coexist, then the moments they exist at coexist. So a single indivisible moment coexists with distinct successive moments.

Textual evidence concerning steadfast objects

There is clear textual evidence that a steadfast object can coexist with a succession. At T 1.2.5.29; SBN 65, Hume specifically discusses a steadfast object regarded at different times, coexisting with "a continual succession of perceptions in our mind." At T 1.4.2.29; SBN 200–1 he says,

> I have already observ'd, that time, in a strict sense, implies succession, and that when we apply its idea to any unchangeable object, 'tis only by a fiction of the imagination, by which the unchangeable object is suppos'd to participate of the changes of the co-existent objects, and in particular of that of our perceptions.

In these places in the text, the steadfast objects are external objects. However, Hume almost always uses the word "object" to encompass perceptions as well, and in other places his assumption that a steadfast object exists is specifically an assumption that a steadfast perception exists. At T 1.2.3.7; SBN 35, Hume discusses having a perception but not successive perceptions. He claims that "A man . . . strongly occupy'd with one thought, is insensible of time." It does not matter how large an interval the thought takes up. If it is just a single thought, then it is not a succession of thoughts. So it has no duration (T 1.2.3.11; SBN 37). And so he has no sense of time. Time is, of course, passing; successions he is insensible of coexist with his thought. As Hume states further down, "Wherever we have no successive perceptions, we have no notion of time, even tho' there be a real succession in the objects" (T 1.2.3.7; SBN 35). Thus, there are intervals of time when we have a perception but not successive perceptions, even when there is succession elsewhere.

That a single perception can coexist with a succession is explicit also at T 1.2.3.7; SBN 35, where Hume says that when a given duration coexists with a succession of many perceptions, it seems to go by more slowly than when it coexists with a succession of few perceptions. He is following Locke in thinking that a succession of objects can have more members than the coexistent succession of perceptions of them. When this happens, a given perception smears together the successive objects it is of. In these cases, some member of one succession coexists with successive members of the other.

There is further evidence that Hume believes in single perceptions that are not brief, but not successions. He thinks that a steadfast object produces "none but coexistent impressions" (T 1.2.3.8; SBN 36), therefore not successive ones.[5] At T 1.4.2.33; SBN 203 he says that in thinking of a steadfast object the mind merely continues an idea in existence as opposed to producing any new one.

> When we fix our thought on any object, and suppose it to continue the same for some time; 'tis evident we suppose the change to lie only in the time, and never exert ourselves to produce any new image or idea of the object. The faculties of the mind repose themselves in a manner, and take no more exercise, than what is necessary to continue that idea, of which we were formerly possest, and which subsists without variation or interruption.

And again he says that in this case the passage of time "distinguishes not itself by a different perception or idea" (T 1.4.2.33; SBN 203).[6] Finally, he speaks of having "one constant and uninterrupted perception" as opposed to a succession of many constant but interrupted perceptions (T 1.4.2.35; SBN 204).

Now it might be objected here that perhaps lengthy perceptions are not necessarily successions of perceptions, but surely they are successions of some sort of temporal parts. Thus they have duration. But Hume specifically relies on the fact that "Every thing that enters the mind" is "in *reality*" a perception (T 1.4.2.7; SBN 190). If temporal parts of perceptions have gotten into the mind, then these parts must themselves be perceptions. So a perception that is not a succession of perceptions is not a succession of anything.

Despite the textual evidence, Stroud contends that Hume couldn't believe in steadfast perceptions: "Of course, there is not any single, identical perception which does remain in existence. If there were, we could get the idea of identity directly from the senses, just from having that perception, and there would be nothing "fictitious" about the idea of identity at all."[7]

Here Stroud does not take Hume at his word in the passage I've quoted from T 1.4.2.33; SBN 203. Stroud's argument depends on the assumption that a steadfast impression would give us the idea of identity directly. However, the assumption is suspect. A steadfast impression would at best give us an idea of unity. "One single object conveys the idea of unity, not that of identity" (T 1.4.2.26; SBN 200). The idea of identity is rather "a medium betwixt unity and number; or more properly speaking, is either of them, according to the view, in which we take it" (T 1.4.2.29; SBN 201). The attempt to integrate unity and number, which are incompatible, is what makes the idea of identity a fiction. The senses alone could not give rise to this mongrel idea, even given steadfast impressions. The vagaries of the imagination are required. So Stroud's attempt to stretch Hume's text for Hume's own good is a misreading.

Stroud assumes, as do, for instance, Green and Bennett, that all temporally single perceptions (both impressions and ideas) are uniformly exceedingly brief in their existence. Call this the Brevity Assumption. They say that perceptions that seem long are really uninterrupted successions of exactly resembling perceptions. Stroud thinks that contemplating a steadfast unchanging object causes a constant series of these briefly existing perceptions, viz., impressions.[8] On this view, the contrast between succession and steadfast object is really a contrast between *perceivable* succession and *unperceivable* succession. One could even say it is the unperceivability of successiveness that prevents getting the idea of time from a steadfast object (see T 1.2.3.7; SBN 35). Although there are reasons in favor of this Brevity Assumption, there are weightier reasons against it.

Price argues that the fact that something might be a series (because it might have been interrupted) proves that it *is* a series, and, second, that nothing lasts longer than a brief instant on pain of contradiction—viz., both having and lacking certain relational properties.[9] But whether the assumption is a good thing to believe, is a question independent of whether Hume believed it. Far from giving any proof that Hume did, Price says Hume did not. He quotes Hume speaking of "one constant and uninterrupted perception".[10] So Price's arguments for the assumption do not justify using it to interpret Hume.

Nor does Hume's text, despite some of his ways of phrasing things. The best support for the Brevity Assumption seems to be Hume's claim in a subordinate clause, that perceptions "succeed each other with an inconceivable rapidity" (T 1.4.6.4; SBN 252). This claim would support the Brevity Assumption only if it meant that always a uniformly high number of perceptions go by in a given amount of time. Only then would they have to be uniformly exceedingly brief. However, Hume thinks instead that "according as his [a man's] perceptions succeed each other with greater or less rapidity, the same duration appears longer or shorter to his imagination" (T 1.2.3.7; SBN 35). So, giving the "inconceivable rapidity" clause a reading that supports the Brevity Assumption conflicts with something else Hume says. Instead, we should read that clause as saying that there is only an inconceivably small gap, if any, between successive perceptions: when one ceases to exist there is no delay before the coming into being of its successor.

Hume does go on to say that the perceptions "are in a perpetual flux and movement." But this is just to say that they "pass, re-pass, glide away, and mingle" in the mind-theater. Actors can do these things without being exceedingly short-lived; so can perceptions. In the same section Hume says "nor is there any single power of the soul, which remains unalterably the same, perhaps for one moment." But here he is not talking about perceptions, which are not powers. He is talking about sight, hearing, imagination, etc., our "senses and faculties," which are constantly subject to some new input, though not necessarily to all new input (T 1.4.6.4; SBN 252–3). Still, I think he exaggerates here. The imagination of the man "strongly occupy'd with one thought" would seem unchanging for that time (T 1.2.3.7; SBN 35).

Hume uses phrases such as "the incessant revolutions, which we are conscious of in ourselves" (T 1.4.2.10; SBN 191). But the Brevity Assumption finds no support here. A republic, as much as a mind, undergoes "incessant changes of its parts" (T 1.4.6.19; SBN 261). The relevant parts of a republic (i.e., its members) are not momentary, so incessant change of parts need not entail brevity of parts.

Presumably Stroud and the others would take the fact that perceptions are "perishing" (T 1.4.2.15; SBN 194) (generalizing from *impressions* which are what are mentioned in Hume's text) to support the Brevity Assumption. But the fact that perceptions are relatively short-lived entails neither that they are momentary nor that all are uniformly so. I assume that for Hume "perishing" is in opposition to "enduring," the way that for us "perishable" is in opposition to "durable." They are relative terms. Stroud seems to quote the phrase "momentary and fleeting," but there is no citation.[11] I have not yet found this phrase in the *Treatise*. The phrase "fleeting and perishing" occurs, but there Hume is clearly referring to a certain subset of "internal impressions"—i.e., those to which we are not tempted to attribute distinct and continued existence—and saying how we *regard* them (T 1.4.2.20; SBN 195). And even so, the phrase does not entail that these impressions are exceedingly brief. It might seem that something perishing is not continued and so is brief. Hume does use "perishing" and "continued" as opposites. But Hume clearly uses "continued" as short for "continued when no longer perceived" (T 1.4.2.3; SBN 188). The opposite of this does not entail brief.

Hume indeed says that "there is no impression constant and invariable." But it is clear from the context that he means there is none constant and invariable "thro' the whole course of our lives" (T 1.4.6.2; SBN 251). He is not asserting here that all are exceedingly brief. Also Hume does say, "'Tis impossible for the mind to fix itself steadily upon one idea for any considerable time" (T 2.1.4.2; SBN 283). But a length of time may well not be a considerable length of time, and yet still be longer than a brief moment. The Brevity Assumption cannot be sustained.

Given the textual evidence in favor of steadfast objects, including steadfast perceptions, it is hard to understand why Waxman in his interesting book would claim that "Hume denied the possibility of an unchanging view of an unchanging object."[12] There is no citation that directly supports attributing this denial to Hume. Waxman does in the next paragraph quote Hume's claim that "all impressions are internal and perishing existences, and appear as such" (T 1.4.2.15; SBN 194). Even if this quotation about impressions generalizes to perceptions, it doesn't entail that all perceptions perish at a uniform rate. Waxman accompanies this quotation with Hume's claim that all "actions and sensations of the mind" are as they appear (T 1.4.2.7; SBN 190). Though supporting the thought that perishing perceptions appear perishing, it cuts against Waxman's earlier argument that if there were steadfast perceptions within the imagination, we wouldn't know it. We wouldn't, Waxman says, because the imagination would be on "freeze-frame" so there would be no sense of time.[13] In that case, however, a perception wouldn't be as it appears, after all. Independent of this point, Waxman's argument that we would never know if we had steadfast perceptions is suspect anyway. Perhaps the movie-camera analogy is the problem. The analogy makes it plausible to assume that there is only a single succession of perceptions in the imagination, just as there is only one reel of film in the camera. But why should we assume this? Let's rather assume something less like a movie camera and more like a mind, in which various successions coexist, successions occurring at relatively different rates so that relatively more steadfast perceptions coexist with relatively less steadfast ones. Thus we can be aware of steadfast perceptions as such.

Waxman might retreat to Locke to oppose this view of mind, but Locke actually supports it. He specifically defines "Contemplation" as an idea "held there [in view] long under attentive consideration," and "Study" as "when the mind with great earnestness, and of choice, fixes its view on any *Idea*, considers it on all sides, and will not be called off by the ordinary sollicitation of other *Ideas*" (*Essay*, 2.19.1). Notice the being held for a long time in the first definition, and the coexistence of the studied single idea with the various presumably successive soliciting ideas in the second definition. Or consider Locke's explanation why things in very slow motion are not perceived to move. It is because "their change of distance is so slow, that it causes no new *Ideas* in us, but a good while one after another: And so not causing a constant train of new *Ideas*, to follow one another immediately in our Minds, we have no Perception of Motion" (*Essay*, 2.14.7). In other words, when an idea of a thing moving very slowly stays in mind a "good while" before being succeeded by another, we do not notice the motion. Think of watching the minute hand of a clock. It seems stationary until suddenly we notice it has moved. At last a new idea of sense has replaced the previous one. Meanwhile our train of impatient thoughts coexisted with the relatively steadfast perception of the minute hand. Otherwise we would not have noticed the passage of time and the slowness of the hand.

Waxman has two further objections to steadfast perceptions, though without argument, so I will have to guess about their details.[14] The first objection is that steadfast perceptions are precluded by Hume's separability principle. The second objection is one raised also by Barry Stroud in conversation some years before—that a steadfast perception would not be perfectly known. The idea behind both seems to be that a steadfast idea would have temporal subdivisions. The first objection assumes that these subdivisions are separable and so the steadfast idea is really many things, not one thing after all. The second objection assumes that the subdivisions could be known only as they come into existence, and only while they are in existence. So the steadfast perception as a whole cannot be perfectly known. The answer to both objections is that there are no temporal subdivisions of steadfast perceptions.[15] They lack duration. Thinking there are subdivisions assumes a theory of time that is not Hume's. Detailing Hume's theory on this point is what I will do next.

Thus, for Hume, there are steadfast objects, including steadfast perceptions, that coexist with successions.

Moments co-existing with successions of moments

Hume may have held this view, but it seems inconsistent: A steadfast object lacks duration because it is not a succession, but would seem to have duration because it exists more than just briefly.

The problem arises because we think that existing more than just briefly means existing at more than one moment (which on the Humean view amounts to being a succession with parts at those moments). However, Hume in effect proposes an alternative: Existing more than just briefly can alternatively mean existing at a single moment that coexists with successive moments. We have trouble making room for Hume's proposal because we think of time as like a line, and moments as its smallest parts.[16] However, in thinking of time and coexistent moments, a diagram such as the following is better, in which time is more like a brick wall:

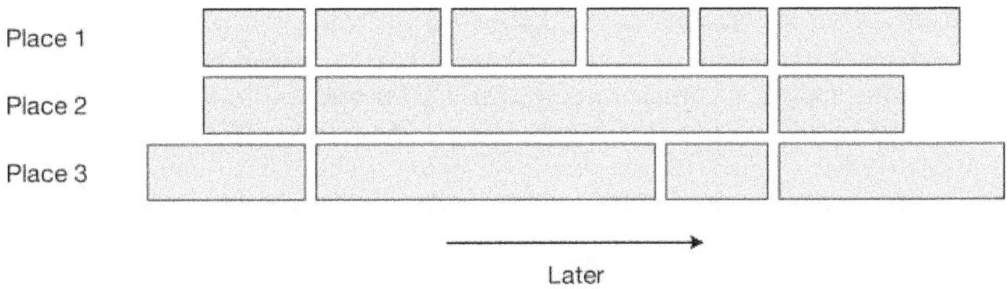

Figure 3.1

The blocks represent moments. Parts of blocks do not represent moments, nor do they represent parts of moments. For Hume, a moment is any thing that is not a succession just insofar as it is part of a succession. That is, moments are abstractions from members of successions. Some moments are abstractions from steadfast objects. Coexistence of moments is represented by the fact that two blocks could be cut by the same vertical line.[17] However, this aspect of the diagram is not supposed to represent anything in reality that explains coexistence. Coexistence is taken to be primitive. At this point, for purposes of resolving the contradiction, I am concerned with coexistence in general (whether past, present, or future) and not just coexistence in the present. That any vertical line which could be drawn through one block would be to the right of any vertical line which could be drawn through another, represents the fact that the first moment is later than the second. Hume thinks that, strictly speaking, only spatial minima are in space, so strictly speaking nothing is in more than one place [T 1.2.2.3; SBN 30–1]. Consequently, in the diagram, no moment is in more than one place.

I have emphasized coexistent successions in different places to draw attention to what is distinctive about Hume's view. However, sameness of place cannot be the only way a succession is united. Hume thinks some successions—particularly of perceptions—are in no place at all, as for instance a succession of tastes or smells or sounds or passions (see T 1.4.5.10–11; SBN 234–7). In fact "the greatest part of beings" exist and yet are nowhere (T 1.4.5.10; SBN 235). Therefore, what is distinctive about Hume's view is precisely that some indivisible moment coexists with distinct successive moments, not that they coexist in different places.

If not always place, then what unifies successions? What makes moments members of the same succession of moments? We experience time by experiencing various coexistent successions of objects. It is the ideas of these that we use to form the abstract idea of time. The experienced unity of successions of objects is a result of the principles of association of ideas (T 1.1.4.1–7; SBN 10–13). In all cases, temporal contiguity plays a role. Sameness of place helps unify some successions. The successions of things in no place, however, must be united by the help of resemblance or causation. So, if Hume enjoys the succession of tastes in a sip of a complex claret, their resemblance as tastes and their having a common cause helps unify the succession. If Hume simultaneously listens to a birdsong and feels a change of mood, each of these two successions of perceptions likewise are unified without appeal to sameness of place. (Though, it must be said, Hume will naturally attribute place to each to "compleat the union" [T 1.4.5.12; SBN 237]—he will hear the song as in the bird's throat and feel the moods as in his own breast.) The abstract idea of time abstracts from the particular natural relations unifying a succession, just retaining their being unified some way or other.

That there are coexistent things that are nowhere, seems inconsistent with Hume's claim at T 1.2.3.8; SBN 36 that "that quality of the co-existence of parts belongs to extension." However, here he must mean that all extended things have coexistent parts. He can't mean that all coexistent things form something extended. Coexistent tastes or smells or sounds or passions do not (T 1.4.5.7–11; SBN 234–7).

I have also pictured successions of moments as not overlapping in the diagram. However, successions of perceptions or objects united by natural relations (see T 1.1.5.1; SBN 13–14) might have some members in common. Suppose the left-hand and right-hand parts of a piano piece come together at the same note for a couple of beats and then separate again. One might hear the successions of lower notes and higher notes merging then diverging. However, this fact about time as it is experienced does not affect the core fact about time, on Hume's view, that an indivisible moment can coexist with successive moments. The above diagram as it stands serves for the current purpose simply of illustrating this fact.

His is a peculiar view, admittedly. Moments can coexist. Some moments are longer than others. The view still feels inconsistent. But that is because we tend to take "moment" to refer to brief, successive, simple temporal locations. We never considered that these characteristics might come apart. Hume uses the word more weakly to refer to simple temporal locations, whether or not brief or successive. To dispel the remaining air of inconsistency, I will formalize Hume's view that some moments coexist with successive moments.[18] The domain of discourse is a set of moments. The primitive relation is the later-than relation, represented by " > ". I will not assume that moments are ordered discretely, though Hume believes they are, since this assumption is irrelevant to the worries about inconsistency.

Assumptions

Later-Than is irreflexive, is asymmetric, and is transitive:

1. (x)~(x > x)
(For all x, it is not the case that x is later than x.)

2. $(x)(y)(x > y \rightarrow \sim(y > x))$
(For all x and all y, if x is later than y then it is not the case that y is later than x.)

3. $(x)(y)(z)((x > y \;\&\; y > z) \rightarrow x > z)$
(For all x, all y, and all z, if x is later than y and y is later than z then x is later than z.)

Definition of Coexists-With:

4. $x@y =_{df.} \sim(x > y) \;\&\; \sim(y > x)$
(By definition, x coexists with y if and only if it is not the case that x is later than y and it is not the case that y is later than x.)

There is at least one moment which some successive moments coexist with:

5. $(\exists x)(\exists y)(\exists z)(x > y \;\&\; (x@z \;\&\; y@z))$
(For some x, some y and some z, x is later than y and x coexists with z and y coexists with z.)

If one moment coexists with another, then any moment later than one is later than any moment the other is later than.

6. $(w)(x)(y)(z)((x > w \;\&\; x@y \;\&\; z > y) \rightarrow z > w)$
(For all w, all x, all y, and all z, if x is later than w and x coexists with y and z is later than y, then z is later than w.)

Consistency

(1)–(6) do not jointly entail a contradiction: one model consists of {1,2,3,4} as the domain, and {<4,3>,<4,1>,<3,1>,<2,1>} as the extension of " > ". In this model, 4 is later than 3, both are coexistent with 2, and all three are later than 1.

Consequences

By (4) and (1), Coexists-With is reflexive. Since nothing is later than itself, by definition everything is coexistent with itself.

7. $(x)(x@x)$
(For all x, x coexists with x.)

By (4) and commutivity, Coexists-With is symmetric. If neither x nor y is later than the other, then each coexists with the other.

8. (x)(y)(x@y → y@x)
(For all x and all y, if x coexists with y then y coexists with x.)

By (4) and (5), Coexists-With is not transitive. Since successive moments can coexist with the same moment, moments that do not coexist can coexist with the same moment.

9.~(x)(y)(z)((x@y & y@z) → x@z)
(It is not the case that for all x, all y, and all z, if x coexists with y and y coexists with z then x coexists with z.)

It follows from (6) and (4) and (1) that a moment between moments which coexist with the same moment, also coexists with that moment. For suppose the in-between moment did not coexist with the steadfast moment its flanking moments coexist with. Then the in-between moment would have to be either earlier or later than the steadfast moment. But then there are coexisting moments (i.e., the steadfast moment and a flanking moment) such that our moment in question is earlier than one and later than the other. This couldn't happen unless it could be later than itself, which it can't be. Thus:

10. (x)(y)(z)((x@z & y@z) → (w)((x > w & w > y) → w@z))
(For all x, all y, and all z, if x coexists with z and y coexists with z, then for all w, if x is later than w and w is later than y then w coexists with z.)

It follows from (6) and (4) that successive moments cannot each coexist with both of successive moments. Otherwise the earlier of the first succession and the later of the second succession would coexist and yet also be in succession. Thus:

11. (x)(y)(z)((x > y & (x@z & y@z)) → ~(∃w)(w > z & (x@w & y@w)))
(For all x, all y, and all z, if x is later than y and x coexists with z and y coexists with z, then it is not the case that for some w, w is later than z and x coexists with w and y coexists with w.)

It follows from (6), (1), and (4) that for moments which coexist, at least one is such that anything later than it is later than both. For, suppose there are coexistent moments such that each has a moment later than it but not later than the other. If not later than the other, then what? Not earlier, because the same moment cannot be later than one of coexistent moments yet earlier than the other. Not coexistent, because then the later moment in one sequence would turn out to be coexistent with a moment it is later than. Thus:

12. (x)(y)(x@y → ((z) (z > x → z > y) V (w)(w > y → w > x)))
(For all x and all y, if x coexists with y then either for all z, if z is later than x then z is later than y, or, for all w, if w is later than y then w is later than x.)

Sentences (6), (10), (11), and (12) convey that there is a rough coordination of different successions, despite the lack of the precise coordination afforded by an equivalence relation of simultaneity.[19] The coordination is in both temporal directions. Hume's view could be formalized in just the same way except substituting for Later-Than, Earlier-Than defined as follows:

13. $x < y =_{df.} y > x$
(By definition, x is earlier than y if and only if y is later than x.)

Additional considerations

I can now be more precise about successions of moments. A succession of moments is several moments such that for any two, one of them is later than the other, and such that for any two, either no moment is between them or any moment between them is one of the several. Thus distinct coexisting moments are not in the same succession, though in principle they could be in different successions which otherwise have all the same moments. A temporal succession coexists with another just in case each moment in one coexists with some moment in the other, and vice versa.

Not only is Hume's view consistent, there are even some advantages of conceiving of time in this Humean way. This conception helps support two deeply held commonsense convictions about time: that time is very unlike space, and that time flows.

As for the first point, something without temporal parts can coexist with a temporal succession. But something without spatial parts cannot be located along a spatial succession. So time is very unlike space.

The closest one could come to a spatial analogue for a steadfast object would be the perception one would get when looking out a window at a pitch-black sky along a visible (perhaps artificially lit) horizon. Since darkness conveys no idea of extension (T 1.2.5.7; SBN 56), the experience would be like seeing a partless black extent, framed by the window, alongside an extended landscape. Strictly speaking, however, in perceiving the sky there is not a perception of blackness; rather there is no perception at all. So, neither at the level of external objects nor at the level of perceptions is there really something unextended stretched out alongside of something extended.[20] So there is no spatial analogue to a steadfast object.

As for the second point, any temporal succession moves with respect to any other it coexists with. For example a succession with five moments moves with respect to a succession with seven moments, at the rate of five moments per seven moments.[21] If we perceivers of external events have a background succession of perceptions[22] with which to compare any sequence of *sense* impressions, then successions of sensed external events will always seem to flow in comparison with this background succession. The background succession would presumably consist mostly of ideas and impressions of reflection, and perhaps impressions of sense that one is hardly attending to. Successions of the sensed external events one is paying attention to, will seem to flow faster if the sequence of sense impressions speeds up relative to the background succession. Likewise, they will seem to flow more slowly if the reverse holds true.

A counterexample seems to be the perceived circle of fire when the coal is wheeling so fast that it travels the whole circuit during the time of our briefest visual impression (T 1.2.3.7; SBN 35). Here, the sensed external succession is very fast relative to the background succession, yet there appears to be something stationary—the circle of fire—not something fast. However, there is a straightforward response. It is the comparison of background succession to succession of sense impressions that matters, not to what is really happening in the world. It may well be that when the visual impressions can't keep up, the mind stops exerting itself to produce new impressions and lets a steadfast one serve (cf. T 1.4.2.33; SBN 203). So there is no speedy succession of sense impression, so the case would not be a counterexample after all. I must acknowledge, however, that even when seeing the stationary circle of fire, we additionally in some way perceive the coal as something indistinct moving quickly. My guess is that we incorporate some knowledge—the knowledge that the circle-of-fire perception is the culmination of gradually less acute perceptions of something gradually moving faster. Plus we see the hand swinging the cord with the coal at the end. Without some knowledge of the context, we would not perceive the speed in any way. There are further hard cases no doubt, but in general if the background succession goes relatively faster then the sensed events seem relatively slower, and if the background goes relatively slower then the sensed events seem relatively faster.

That all coexistent successions flow relative to each other does not preclude us from fixing on an intersubjective standard for the rate of flow of successions. We can take some external successions, like the tickings of clocks, to be the standard by which we judge the rate of other successions. If we come to believe, as we tend to, that there is an ultimate succession of the briefest possible moments that provides the standard for all flow, then philosophers driven beyond common sense may even come to think that the standard succession itself does not flow.[23]

Objections

An objection seems to arise when we shift from thinking of coexistence in general to coexistence in the present. Suppose a longer moment coexists with successive moments, and suppose that one of these successive moments is present and another past. Then the longer moment is partly present and partly past. But anything past is not present. So the longer moment is itself a succession of distinct moments—one present, one past—on pain of contradiction.

However, this objection assumes that anything that coexists with a past moment is partly past. By denying that steadfast objects and the moments they occupy have parts, despite coexisting with successions, Hume is committed to denying this assumption. The assumption stems from a view in which presentness is a global phenomenon and moments go in and out of presentness in lockstep. For Hume, on the other hand, presentness would be a local phenomenon—local to particular successions. Moments would go in and out of presentness more raggedly, the only constraint being that all present moments coexist with each other.

Another objection to my interpretation of Hume, as interpretation, could be based on his claim that distinct moments cannot be coexistent, since it is of the essence of time that all its parts be successive (T 1.2.2.4; SBN 31). He later says, "'Tis also evident, that these parts are not co-existent: For that quality of the co-existence of parts belongs to extension, and is what distinguishes it from duration" (T 1.2.3.8; SBN 36). And again, "On the contrary, time or succession, tho' it consists likewise of parts, never presents to us more than one at once; nor is it possible for any two of them ever to be co-existent" (T 2.3.7.5; SBN 429). In all three places Hume seems clear that moments can't coexist in time.

However, the opposite seems entailed by Hume's claims that commit him to steadfast objects, as we have seen.

The best way to reconcile all these claims consists of two parts. The first is to realize that Hume is using "time" and "duration" interchangeably with "succession" (see T 1.2.3.11; SBN 37). When he talks about time or duration, he is talking about being a succession. If this is so, then it makes sense for him to say that having parts in succession is the essence of time or duration, for it is the essence of being a succession and is to be contrasted with having parts that are coexistent and not in succession.

The second part of the reconciliation is to conceive of the manner in which more than one coexistent succession is arrayed. Hume does not supply a term for this but it could be "succession-coexistent-with-succession", or less awkwardly, "co-duration". Thus, although Hume does not allow any moments of a duration to coexist, he does allow some moments of a co-duration to coexist.[24] Once co-duration has been recognized, it seems that it could be a legitimate extended meaning for the word "time." However, Hume doesn't extend the meaning because recognizing co-duration requires unnatural care and attention. What is natural is to attribute duration to steadfast objects and so be unable to see the need for, or even the possibility of, coexisting moments.

Fictitious duration

Despite the fact that steadfast objects lack duration, we naturally come to think of steadfast objects as having duration—that is, as being successions—because of the coexisting successiveness of perceptions in our minds

(T 1.2.5.29; SBN 65). It happens by the psychological mechanism of confounding: "we may in general observe, that wherever the actions of the mind in forming any two ideas are the same or resembling, we are very apt to confound these ideas, and take the one for the other" (T 1.2.5.21; SBN 61).

In other words, the mind presents:

> other related ideas in lieu of that, which the mind desir'd at first to survey. This change we are not always sensible of; but continuing still the same train of thought, make use of the related idea, which is presented to us, and employ it in our reasoning, as if it were the same with what we demanded.
>
> (T 1.2.5.20; SBN 61)

When we are not aware that the wrong idea has been substituted, we continue thinking as if no substitution had been made and thinking that the characteristics presented by the substitute idea apply to what we began thinking about. What is at issue here is how we come to think of an object which is in fact steadfast, as if it rather had duration. In other words, the issue is how we come to think of something which is in fact a single thing coexisting with a succession, as if it rather were many things in succession. In doing so we lose sight of its steadfastness.

Continuing to attend to its steadfastness will come later in the order of acquisition of ideas when we acquire the idea of identity (T 1.4.2.26–30; SBN 200–1). By then we have acquired the tenacious habit of regarding everything that coexists with a succession as having duration. At some point, we then fully appreciate the steadfastness of a steadfast object perceived as such. We do not instantly confound the idea with an idea of a succession, as before. Thinking of the object as steadfast, we think of it as one. Yet the strong habit reasserts itself. Thinking of the object as having duration, we think of it as many. The idea of identity arises when we realize that we are thinking of the steadfast object as both one and many. Afterwards, we start applying the idea of identity to what are really successions when acquiring the ideas of body and of self (T 1.4.2.31–7, 1.4.6.5–16; SBN 202–5, 253–60). Even later we distinguish bodies from perceptions (T 1.4.2.44–6; SBN 210–11). Only after this would we be able to attribute duration to steadfast perceptions. Presumably we would do so by attributing identity to them, since their steadfastness could not be overlooked if we are paying attention to them.

Again, however, all this will be later in the order of acquisition. Here I focus just on the first step when we "confound" the idea of a steadfast object with an idea of a succession. Unawares, we take what is in fact steadfast to be successive. That is, we start with an idea of a steadfast object and fail to notice that a resembling idea of something with duration gets inadvertently substituted. Hume doesn't say so, but it is likely that the substituting idea is of a succession of things exactly resembling the steadfast object (except with respect to temporal length).[25] After all, there is no change between the steadfast object's appearances.

The explanation for the substitution occurs at T 1.2.5.29; SBN 65. He says that "there is a continual succession of perceptions in our mind; so that the idea of time" is "ever present with us." Consequently, "we are apt to apply" the idea of time to a steadfast object that we pay attention to at different times, but not continuously.[26] For instance "we consider a stedfast object at five-a-clock, and regard the same at six." When we do so, "the first and second appearances of the object, being compar'd with the succession of our perceptions, seem equally remov'd as if the object had really chang'd." What he means here, I contend, is that these successive experiences of a single steadfast object seem just like experience of a succession of objects that are not steadfast. Consider Hume's phrase "seem equally remov'd." Johnson's first definition of "Remove" is "To put from its place; to take or put away." Recall that a steadfast object is one fixed in place. So an object that is removed is not steadfast; it is rather one member of a succession; it is changed in the sense of replaced. Thus, because we are constantly experiencing succession, we tend to think of an object experienced at one time (i.e., in the first appearance) and that object experienced at a later time (i.e., in the second appearance), not as one steadfast thing but as many things in succession. The fact is, though, that the object *is* steadfast. So the idea of it as a succession is a fiction.

Think of the fiction this way: Instead of (i) coexisting with all members of a succession, we rather attribute to the steadfast object (ii) successively coexisting with each member to the exclusion of the others. The former is how a steadfast object coexists with a succession. The latter is how several things in succession coexist with another succession: The first member of one succession coexists with just the first member of the other, then the second coexists with just the second, then the third coexists with just the third, etc. In this way we think of the steadfast object as a succession.

Thus, we apply the idea of duration to what is in fact a steadfast object "in the same manner as if every moment were distinguish'd by a different position, or an alteration of the object." We are reinforced in this fiction by our recognition of two additional things. First, the object *could have* changed between five and six o'clock. It could have been replaced, moved, or altered. So the potential for change puts one in mind of the change in a duration. The second is harder to interpret. Hume tells us that experience also shows us "that the unchangeable or rather fictitious duration has the same effect upon every quality, by encreasing or diminishing it, as that succession, which is obvious to the senses." He cannot be talking about the qualities of the steadfast object. If they changed, then the duration would be real even if imperceptible. He rather says the duration is fictitious. So Hume must be talking of the qualities of coexistent objects. Things coexisting with steadfast objects alter just as much as when coexisting with successions of objects. For example, when one returns one's gaze to the fire after contemplating the unchanging view out the window for a while, the fire is perceived to have burned down as much as if one had been continuously watching the clock (T 1.2.5.29; SBN 65).

As a result, where we ought to have an idea of the steadfast object as steadfast, we end up having an idea of it as a succession. Hume says the result is a fictional "idea of time without a changeable existence" (T 1.2.5.29; SBN 65). More perspicuously it is an idea of duration being used when we fancy that we are continuing to think about something steadfast.

This interpretation relies on Hume's claim that only successions have duration (T 1.2.3.11; SBN 37). However two interpretive problems arise. First, when he says in the passage quoted above "as if every moment were distinguished by a different position," he seems to imply that the idea of duration would properly be applied to something just in virtue of its moving. However, he has elsewhere precluded that propriety (T 1.3.1.1, 1.4.6.8; SBN 69, 255). He must, rather, mean that the idea of duration is improperly applied to the steadfast object in the same manner as it is properly applied to the succession of places occupied by a moving object. Second, Hume, when he continues with "or an alteration of the object," seems to suggest that a non-succession—namely, something altering—could have duration, contrary to what I have argued. However, Hume regards altering things as really successions of distinct things. He regards "variation" as "evidently contrary" to identity through time (T 1.4.3.2; SBN 219). Presumably what he has in mind is the following: Variation amounts to having, then lacking, some (nonrelational) quality. If what has the quality is numerically identical with what lacks it, then a contradiction is true of that thing. So they are not identical. In other words, Hume is assuming a powerful and tempting solution to the standard problem of alteration.[27]

So, again, we get into the habit of applying the idea of duration to everything, whether a succession of changeable objects or not.

This fiction, coupled with a tendency to believe in exact standards for judging equality, explains why it is so hard to conceive of a single moment coexisting with a succession of moments, and so is the source of the failure to distinguish duration from co-duration. Let me supply a Humean account of the failure. We naturally think of the steadfast object as a succession. So we naturally think that the steadfast object occupies a succession of moments. Now, suppose a coexisting succession of objects begins, say, a little before the steadfast object (i.e., the first member of the succession also coexists with something earlier than the steadfast object). We can't yet think that the steadfast object and the first member of the coexisting succession occupy the exact same moment. We notice, however, that a fast, long-term succession would provide sub-successions that could very closely approximate the feigned "duration" of any steadfast object or the duration of any succession. Noticing this, the mind tends to concoct a fiction that there is an exact standard for judging the equality of the duration of any two "things" (whether steadfast object or succession). This standard for time is exactly analogous to the one for space of which Hume says, "But tho' this standard be only imaginary, the fiction however is very natural" (T 1.2.4.24; SBN 48). He goes on to say:

> This appears very conspicuously with regard to time; where tho' 'tis evident we have no exact method of determining the proportions of parts, not even so exact as in extension, yet the various corrections of our measures, and their different degrees of exactness, have given us an obscure and implicit notion of a perfect and entire equality.
>
> (T 1.2.4.24; SBN 48)

This exact standard would, presumably, be a succession none of whose single moments coexist with any succession. They would be exceedingly brief. If we think every other succession occupies a raft of exceedingly brief moments exactly coexistent with these, then there is no longer any reason to distinguish coexistent moments. None are longer than others. The succession of moments at any one place would exactly resemble the succession at any other place, as well as those in no place. So difference in place, or in any other principle of unity for a succession, would seem irrelevant to succession of moments. So any distinction between duration and co-duration would seem a distinction without a difference.

So we naturally come to think that a steadfast object occupies a succession of exceedingly brief moments, and indeed occupies the same succession of exceedingly brief moments that anything coexisting with it occupies. So it is unnatural to think of a single partless moment coexisting with a succession of moments.

When we habitually think of time uniformly as a succession of brief moments, and habitually think of anything in time, when not brief, to have duration, then the stage is set to acquire the idea of identity.[28]

4 Identity

Hume's discussion of identity is part of his explanation how we come to believe in objects that continue to exist unperceived. He means to show that "the *invariableness* and *uninterruptedness* of any object, thro' a suppos'd variation in time" are essential to identity through time (T 1.4.2.30; SBN 201). In other words, necessarily if something is identical through time then it does not alter and there are no gaps in its existence and it exists at distinct times. He later confirms that each of the first two are essential by saying that that variation is "evidently contrary" to identity (T 1.4.3.2; SBN 219), and "'tis a false opinion that any of our objects, or perceptions, are identically the same after an interruption [in existence as opposed to an interruption just in appearance]" (T 1.4.2.43; SBN 209). He earlier said that identity "is common to every being, whose existence has any duration" (T 1.1.5.2; SBN 14), though it is there too soon for him to note that strictly speaking anything with duration is some beings, not a being, and that a being with duration is the fiction underlying the idea of identity (a fiction that the word "suppos'd" reflects). After the discussion of identity, he goes on to show how successive interrupted perceptions with the first quality—invariableness (or "constancy")—come falsely to be attributed the second quality—uninterruptedness—and then how belief in something having continued existence unperceived, is a compromise between the attributed uninterruptedness and the manifest interruptedness of these perceptions.

He takes the first two qualities to be essential to identity, because the experience of something with these qualities—namely, a steadfast object—is necessary for acquiring the idea of identity. Only such an experience can cause the idea of identity, and then only with the prior conviction that anything which does not exist briefly, endures (i.e., has the third quality). That the fiction of a steadfast object with duration should be the idea of identity is alien to us nowadays and seems somewhat ridiculous. Surprisingly, Hume has an important reason in its favor. He raises a difficulty with conceiving of identity, and the fiction of a steadfast yet enduring object seems well suited for trying to get around the difficulty.

The difficulty is raised by the clear fact that identity is something we can be unsure about. We are able to think of two things while leaving it open whether or not they are identical. Their identity or distinctness "is never discoverable merely from their ideas."[1] Identity is a relation "such as may be chang'd without any change in the ideas" of the relata (T 1.3.1.1; SBN 69). So, the concept of identity must be such that we are able to represent the two things as perhaps identical and yet represent the very same things as perhaps distinct. Thus, we must be able to see them as one from one point of view and see them as many from another. As Hume puts it, the idea of identity must be an idea of "a medium betwixt unity and number" (T 1.4.2.29; SBN 201). This difficulty is what motivates Hume to give the principle of identity as steadfastness yet with duration. When we experience something's steadfastness while still attributing duration to it, we get a fictional idea that serves well enough as an idea of it as one and yet again as many. As a steadfast object it is one, and as having duration it is many. The idea is at root contradictory and confused, but it seems to Hume to be the only one we have that will do.

The difficulty Hume raises can be hard to distinguish from Frege's famous puzzle concerning the truth and informativeness of statements of identity.[2] As Bennett says, "I invite you to work through the text and then agree that at best it comes down to the following. Hume is contending that there can be no informative, a posteriori, true identity statements, because 'x is x' is always trivial (unity), and 'x is y' is always false (number)."[3] It is hard not to agree. Bennett's interpretation brings much of what Hume says into apparent focus and allows us in good conscience to dismiss as confused what remains blurry. Nonetheless, Hume is up to something else. In order to begin to distinguish Hume's difficulty from Frege's puzzle, we have to take seriously something we are inclined not to, namely, Hume's claim to the effect that a sentence like "Hume is identical with Hume" does not give a case of identity. We tend to overlook or discount this claim because we can't make sense of it, and instead read Hume as simply but clumsily making the Fregean point that such sentences are not informative. Influenced by Frege, we think that a sentence such as "Hume is identical with Hume" gives us a paradigm case of identity. Hume, however, would consider it a degenerate case. It is a degenerate case of identity analogous to the way in which a point is a degenerate case of a circle. A point can be thought of as a circle with zero radius. Is the claim "A point is a circle" true? One may be inclined to say yes. It is true that all points on the perimeter (such as it is) are equidistant from a central point. However one may just as well be inclined to say no. The fact that the distance is zero makes the case odd and uncharacteristic. One wouldn't mark a point on a piece of paper in order to give someone else an idea of what a circle is.

Analogously, Hume thinks that a claim such as "Hume is identical with Hume" gives a case that is odd and uncharacteristic. For Hume, the relation of identity standardly holds between things that differ in some way. Only concerning such things can the question of their identity arise. Only then can we begin to imagine that the very things we are thinking of might have turned out to be distinct. When there is patently a single thing that cannot be imagined to be distinct from itself, there is no identity or at best there is degenerate identity. Now the only way Hume can think of for something to differ from itself is for it to have changed. The most fundamental way of changing is simply to change temporal location as time passes. All other changes depend on this one. Thus Hume quickly turns his discussion of identity into a discussion of identity through time.

We are likely to become impatient with Hume's obtuseness here. Obviously nothing can differ from itself. Leibniz's Law makes it clear that everything must resemble itself in all ways on pain of contradiction. The only differing we can get in a case of identity is between different referring expressions or different ways of referring. There is no difference between the morning star and the evening star; there is only the difference between "the morning star" and "the evening star', or between their senses. There is really no question about the identity of the morning star and the evening star. There is at root only question about whether or not there is just one thing that both "morning star" and "evening star" refer to. We cannot literally imagine that the morning star and the evening star are distinct. We can at best imagine that "morning star" refers to one thing and "evening star" refers to another.

Our impatience is too hasty, however, if we want to understand and perhaps learn. Certainly, there is some difference between something before it has changed and the same thing after it has changed. Such a difference is not simply a matter of difference in referring expression. Change is not simply a matter of how we talk. Whatever sort of difference there is between something before it has changed and the same thing after it has changed, must ultimately be consistent with Leibniz's Law. How exactly the difference is consistent is a matter of much dispute, and we would be rash to say we understand it.[4] Thus, there is some way in which something can differ from itself, ultimately consistent with Leibniz's Law and not just a matter of difference in referring expressions. Such a difference, Hume thinks, must be present in a standard case of identity. Without it the case is degenerate.

Fregeans likely get their inspiration from arithmetic equality. Equality is their model for identity. The trouble is, numbers as treated in arithmetic perch imperturbably in eternity; issues of change do not arise. Hume, on the other hand, is concerned with the change-filled world of sense. Any length of time experienced as such involves the experience of change. Keeping track of things which in fact do survive change, though might not have, becomes an important task. It is here that we employ the idea of identity. Identity is standardly identity through change.

I have been talking as if Hume believes that there is identity through change. Of course he does, when thinking commonsensically, and his goal is to explain how the idea of identity becomes part of one's commonsense way of thinking about the world. However, when thinking metaphysically, Hume sees such alteration as a fiction. As he says when discussing the ancient metaphysics of substance, variation is “evidently contrary” to identity (T 1.4.3.2; SBN 219). Metaphysically, Hume makes a severe application of Leibniz’s Law. So, explaining how we come to have the concept of identity is at root explaining how we come to construct the fiction of identity through change.[5]

Most cases of this fiction result from confusing closely related qualities (or objects) in succession, with a case of identity (T 1.4.2.31–6, 1.4.6.5–16; SBN 201–5, 253–60).[6] So, most cases of the fiction of identity involve a core fiction of identity. The core fiction is inspired by the clearest case of what is commonsensically regarded as identity through change—where the change lies only in the time (cf. T 1.4.2.33; SBN 203). The object undergoes no alteration, no replacement, no change except for existing at a later time than previously. This difference is the only difference between it at the earlier time and itself at the later time. That there is just one object, not two, is as clear in this case as it can be in any. The more alterations there are, the less clear that there is one object as opposed to a succession of different objects. Yet the mere difference in time is enough to open up the conceivability of two objects being involved, rather than just one. Such a case is Hume’s paradigm case of identity. The steadfast object at the earlier time is identical with itself at the later. Thus identity is standardly identity through change which, at its core, is identity through time.

That we can think of two things and leave it open whether or not they are identical is needed for being able to conceive of their identity, according to Hume. We must be able to represent them as being one and the same thing and yet be able to represent them—the same them—as being two distinct things. If we are only able to represent them as one, then we are not representing identity with its requisite difference. If we are only able to represent them as two, then we are not representing identity with its requisite oneness. In the presentation of his difficulty, Hume shows why these two unsatisfactory options are, apparently, the only options.

To see the problem, we must realize that Hume’s concern is with intentional objects. To understand his concern we must distinguish between (i) what there is which an idea represents, and (ii) what an idea represents there as being.[7] The phrasing tries to reflect the scope distinction Quine uses to capture the difference between “relational” and “notional” representing. The contrast is whether “represent” has short or long scope with respect to “there is.”[8] This distinction is not one that Hume explicitly makes, nor is careful about, but it is needed for proper interpretation of Hume and anyone employing the theory of ideas. The distinction would be familiar from, for instance, Descartes’s discussion of “the mode of being by which a thing exists objectively < or representatively > in the intellect by way of an idea”[9] On this way of thinking, objects of ideas are of two different sorts. Objects of the first sort have real existence. They are represented by ideas because they cause the ideas and the ideas resemble them in some respects. Objects of the second sort have only “intentional inexistence” to use the scholastic phrase resurrected by Brentano.[10] They are the objects portrayed by the ideas independently of whether the ideas are caused by and resemble real world objects. For lack of standard terminology I will call the first sort "intended objects" and the second "intentional objects."[11]

My awkward locutions capture a distinction often implicit in discussions of representation, yet rarely made explicit.[12] Often a distinction is made between the physical properties of the objects we represent and the categories we class them under, that is, what we take them to be.[13] But such a distinction doesn't handle cases in which the number of objects we represent differs from the number of objects we represent there as being. There may be no intended object yet one intentional object. For instance, take a case of "amodal completion" in which a subject represents there as being a triangle when seeing three close circles each with a wedge removed just where a vertex of the triangle would be, even though there in fact is no triangle there to be represented.[14] Similarly, there may be many intended objects yet only one intentional object. In a homemade flip movie we represent there as being a single moving dot, when what there is which we represent is a succession of dots in slightly different positions on a succession of pages.

The closest thing to standard terminology for the distinction I am making, is Ayers' distinction between "real object" and "intentional object of thought." He characterizes the former as "the thing as it exists in reality" and the latter as "the thing as it exists in the mind."[15] This way of putting the distinction might make it seem like a distinction between the properties something really has and the properties someone thinks or perceives it to have. It is like the familiar use of "as" in claims such as "In this optical illusion, people see the equal lines as unequal." However, this distinction between the real and apparent properties of something is not all Ayers means. There might be a case, such as that of a chimera, in which there is nothing for which we can distinguish its real from apparent properties. In such a case the mind does not merely represent some real thing as being a certain way. There is no real thing. Rather, the mind represents there as being something that is a certain way. When we cannot start with something existing and go on to represent it as having certain properties, we have to represent there as being (existing) something with certain properties. When Ayers considers the possibility that there is a real distinction between the "intentional object of thought" and the "real object," he is considering the possibility that the intentional object of thought is not the real object itself, conceived differently than it really is, but rather is entirely something that we represent there as being. Hence my strange locution is needed to more fully capture what Ayers means.

I contrast *what an idea represents there as being* with *what there is which an idea represents.* Simply calling this latter "the thing as it exists in reality" is not as helpful as my phrase, because I am concerned with it insofar as it is an object being thought about or perceived. "Object of thought" is ambiguous between intentional object and intended object, but they are both objects of thought in the respective senses of that phrase. Ayers seems at first to reserve phrases such as "insofar as it is thought of" for "intentional objects of thought," but later allows room for the kind of point I am making here. He mentions "the real sun regarded as extrinsically related to a thought or mind." If I understand him, this is the sort of thing that I mean to connote with "intended object." In contrast, an intentional object is intrinsically related to a thought or mind. It cannot exist without a representing.

My phrase for intentional objects is helpful not just for talking about representing existence, but also for talking about representing identity and distinctness. If we represent what is really one thing as two distinct things, we represent there as being two things. If we represent what are really two things as one and the same thing, we represent there as being one thing.

Much of Hume's talk of the objects of ideas is talk of what the ideas represent there as being. For instance, the association of ideas depends on qualities of their objects (T 1.1.4.1–5; SBN 10–12).[16] But even ideas of imaginary things can be associated. So it is the qualities of what the ideas represent there as being, that are relevant. Or for instance, in his discussion of causation Hume makes a sharp distinction, though he doesn't put it this way, between what there is which we represent and what we ordinarily represent there as being. The former is a constant conjunction or a habitual association. The latter is a necessary connection (T 1.3.14.25–31; SBN 167–70).

Hume's discussion of the principle of identity is not really concerned with the number of intended objects and is centrally concerned with the number of intentional objects—how many objects the ideas represent there as being. Granted, he talks in a misleading way. However, talking about certain characteristics of the intended object and assuming they are accurately represented, is an easy and common way of talking about the characteristics of the intentional object.

Perhaps an analogy will help to see the distinction. Consider an actor portraying a character based on an historical figure, for instance Richard Burbage portraying Shakespeare's Richard III based on the real Richard III. The actor is analogous to an idea, the character to an intentional object and the historical figure to an intended object.[17]

The analogy of the stage can also help illuminate the problem Hume raises. Suppose, as we normally do, that the number of actors on stage represents the number of characters in the dramatic situation. A person might well wonder how it would be possible to stage a situation in which two characters are portrayed as identical but in a way that would have allowed them equally well to have been portrayed as distinct. Imagine a director trying to stage a play of the following form: Scene 1: A man and a man enter the scene in a way that leaves the audience wondering whether or not they are the same man. Scene 2: It turns out that they are the same man. The problem is how many actors to use in Scene 1. If the director uses one actor, what is there for the audience to wonder about? The man and the man are obviously the same man. Furthermore Scene 2 becomes impossible to stage. Likewise, if he uses two actors, the man and the man are obviously distinct men, and again Scene 2 becomes impossible to stage. But what other choice is there than between using one actor or using more than one? None, so there would seem to be no way it could be open to question whether or not those characters are identical. Analogously, Hume's difficulty is that there is no way to represent there as being two things that are perhaps identical and perhaps distinct.

The stage analogy also helps explain why Hume appeals to time to help solve this problem concerning identity. The natural solution for the director is to use one actor and draw attention to him at different times. Perhaps he makes two entrances in Scene 1, without it being clear whether or not he left the stage after the first entrance. Likewise, Hume tries to address the difficulty by showing how we can distinguish an object at one time from itself at another. The analogy is imperfect, however, because the audience for the actor already has the idea of identity, whereas the "audience" for the idea of the steadfast object with duration, is just about to acquire it.

Even without the obstacles of Fregean preconceptions, of Hume's thinking at different levels both commonsensical and metaphysical, and of an unfamiliar distinction between intentional and intended object, Hume's discussion would be hard to understand. It is so compressed and idiosyncratic. Quick summaries with a lot of paraphrase, such as one finds in Kemp Smith, are of no help.[18] The text must be examined step by step.

The difficulty

Hume will make much of the contrast between a single object and more than one object. It is important to note that his main focus will be singleness or multiplicity through time—i.e., steadfastness or duration—and he will be metaphysically rigorous about these. In contrast, he will be merely commonsensical about singleness or multiplicity at a time. Strictly speaking, an ordinary object such as a peach or a melon is a whole composed of many qualities, so strictly speaking it is a multiplicity (T 1.4.3.2, 1.4.3.5; SBN 219, 221). But since that makes no difference to his argument, Hume will follow our natural inclination to speak of it as a single thing.[19]

Hume begins by saying that "the view of any one object is not sufficient to convey the idea of identity" (T 1.4.2.26; SBN 200). This sentence cannot be taken simply at face value because his account will be that the view of a steadfast object *is* sufficient to convey the idea of identity, given proper conceptual preparation. A steadfast object is as single an object as any. Hume also says, "One single object conveys the idea of unity, not that of identity." This isn't quite right either. A single object seen twice could convey the idea of multiplicity, not unity, to someone who didn't know it was the same one. So what Hume must have in mind is that a single object, recognized simply as such, does not convey the idea of identity, but only unity. However, the singleness of the object itself is not really doing any work. An idea of unity could be conveyed even by a multiplicity apprehended inadequately as a single thing. Thus, how many objects there really are that are causing the idea is not what is important at all. What is important is how many objects the idea represents there as being. Thus Hume's concern is with an object represented as being a single thing.

Hume's argument that an object represented as being a single thing cannot convey an idea of identity is this:

> For in that proposition, *an object is the same with itself*, if the idea express'd by the word, *object*, were in no ways distinguish'd from that meant by *itself*; we really shou'd mean nothing, nor wou'd the proposition contain a predicate and a subject, which however are imply'd in this affirmation.
>
> (T 1.4.2.26; SBN 200)

To untangle this passage we need to understand (1) what ideas have to do with propositions, (2) why Hume has suddenly switched from talking of a single object to talking of a single idea, and (3) in what sense the proposition would "mean nothing" if only one idea is involved. An earlier footnote concerning judgments and propositions helps in answering these questions. There he explains that "*conception*, *judgment*, and *reasoning*" are all just "particular ways of conceiving our objects."

First, by "proposition" he is talking of something that *consists of* ideas.

> For *first*, 'tis far from being true, that in every judgment, which we form, we unite two different ideas; since in that proposition, *God is*, or indeed any other, which regards existence, the idea of existence is no distinct idea, which we unite with that of the object, and which is capable of forming a compound idea by the union. *Secondly*, as we can thus form a proposition, which contains only one idea, so we may exert our reason without employing more than two ideas, and without having recourse to a third to serve as a medium betwixt them.
>
> (T 1.3.7.5 n. 20; SBN 96 n. 1)

Here he talks interchangeably of forming a judgment and forming a proposition. Also he makes it clear that a proposition can consist of two or more ideas, as was generally thought at the time, or simply one idea. In the former case, one idea is the subject and the other is the predicate.

It is hard nowadays to understand how a proposition could consist of ideas. After all, ideas for Hume are not like words or sentences; they are images (T 1.1.1.1, 1.1.7.6; SBN 1, 20). However, think of ideas as like actors in the mind that, generally, portray objects and events outside the mind (cf. T 1.4.6.4; SBN 253). By having certain characteristics, ideas represent there as being objects with those characteristics, just as by being stout an actor represents there as being a man—Falstaff, say—who is stout. In this way ideas are propositional.[20]

It is important to note that there is another way for an idea to have a characteristic than to be an image of something with that characteristic. The idea could also be united with another image, e.g., an image of Falstaff united with an image of stoutness. The latter image would be an abstract idea, e.g., an image of some arbitrary stout person viewed in the light of his stoutness. In fact, united ideas are generally structured in this way in contrast to impressions. "Ideas never admit of a total union, but are endow'd with a kind of impenetrability, by which they exclude each other, and are capable of forming a compound by their conjunction, not by their mixture" (T 2.2.6.1; SBN 366).[21]

Hume's difficulty concerns which facts about ideas would represent there as being two things that are identical. In particular, he is concerned with how many ideas it takes to do this. For as the footnote, second, helps us see, Hume connects the number of ideas with the number of objects thought about. It is clear there, that when he talks of some judgments as joining two ideas, he also means that such judgments are considering together two objects. When he talks of some propositions as consisting of just one idea he also means that such a proposition is the considering of just one object. Thus, the quick move at T 1.4.2.26; SBN 200 from talk of a single object to talk of a single idea is no accident.

The connection between the number of ideas and the number of objects is as follows: Hume takes the number of ideas to determine the number of objects the ideas represent there as being. The shifts in T 1.4.2.26; SBN 200 and in the note to T 1.3.7.5; SBN 96 are evidence. Further evidence occurs at T 1.2.1.5; SBN 28. There he says that an image "perfectly simple and indivisible" will "represent as minute and uncompounded what is really great and compos'd of a vast number of parts." In other words, an intended object with many parts can, due to our lack of acuity, cause an idea with no parts which then represents there as being something with no parts. The intended object is a multiplicity of parts: "extension is always a number" (T 1.2.2.3; SBN 30). However, the single, simple idea represents there as being just a single, simple thing. One idea, one intentional object. On the other hand, if an idea with an extended intentional object "agrees to" its object—that is, resembles it—then the idea must represent there as being many things, not just one. For as Hume says, "To say the idea of extension agrees to any thing, is to say it is extended" (T 1.4.5.15; SBN 240). Thus, to represent there as being many things it must itself be many things, since, again, "extension is always a number."[22]

Thus, it is part of Hume's theory of representation that the number of ideas determines the number of objects represented as being. So, again, when he talks of a proposition as consisting of just one idea, he also means that such a proposition is the considering of just one object.

It might seem that there is another way to think of something as one thing: by joining the idea of the thing with the separate idea of unity.

However, this is not an option. Just as there can be no separate idea of existence, so there can be no separate idea of unity, singleness. First, "existence in itself belongs only to unity" (T 1.2.2.3; SBN 30). So it is not possible for something to exist that is not a single thing. So, we cannot even conceive of something existing that is not a unity. For "nothing of which we can form a clear and distinct idea is absurd and impossible" (T 1.1.7.6; SBN 19–20). So, the idea of existence and the idea of unity are inseparably joined. So they are the same idea, since distinct ideas are separable (T 1.1.7.17; SBN 24). But "the idea of existence ... is the very same with the idea of what we conceive to be existent," because nothing can be conceived except as existent (T 1.2.6.2–4; SBN 66–7). So the idea of unity is likewise the very same with the idea of what we conceive to be existent and unitary. There is no separate idea of unity to join with the idea of what we are conceiving. To conceive of something as unitary requires a single idea, and with a single idea we conceive of something as unitary.

So, in the passage in question from T 1.4.2.26; SBN 200, Hume claims that if the proposition, *an object is the same with itself*, contained just one idea and so concerned just one object, then in employing it we would "mean nothing." What can he mean by this phrase? This is the third thing we needed to understand. He cannot mean that we would be saying something meaningless. Employing a proposition consisting of just one idea cannot be sufficient for saying something meaningless. After all, some propositions we can employ meaningfully consist of just one idea—for example the proposition, *God is* (T 1.3.7.5, n. 20; SBN 96, n.1).

Likely, when we "mean nothing" we are employing a proposition that is "trifling," as Locke puts it in Book 4, Chapter 8, "Of Trifling Propositions." One species of such propositions are those in which "we affirm the same Term of it self" (*Essay*, 4.8.2)—for instance "a Chimæra is a Chimæra" (*Essay*, 4.8.3). Such propositions "add no Light to our Understandings, bring no increase to our Knowledge" (*Essay*, 4.8.1). An ignorant person could make a million such claims "of whose truth he may be infallibly certain, and yet not know one thing in the World thereby" (*Essay*, 4.8.3). In general, a trifling proposition is one in which the proposition as a whole adds nothing to the idea that is the subject. There is no predicate idea in addition to the subject idea.

By this criterion, the proposition, *God is*, would be trifling according to Hume. "To reflect on any thing simply, and to reflect on it as existent, are nothing different from each other. That idea, when conjoin'd with the idea of any object, makes no addition to it" (T 1.2.6.4; SBN 66–7). Likewise, the proposition, *God is a single thing*, would be trifling, since it is equivalent. In fact, any proposition containing just one idea would be trifling because the proposition as a whole adds nothing to the idea that it the subject.

It might seem odd for Hume to regard the proposition *God is* as trifling, since there has been some controversy over God's existence. However, Hume's claim is actually a modest one. Basically, he is saying that one cannot conceive of God, or anything for that matter, except as a being. Conceiving of something as a being, in no way entails that it is a being. Just as "a Chimæra is a Chimæra" is only known to be trivially true because it does not commit to there really being one, so "God is" on Hume's construal is known to be trivially true because it does not commit to there really being one. The nontrifling proposition would be "God exists externally", which likely would be to conceive God in spatial or temporal or causal relations to bodies conceived to exist unperceived (T 1.1.5.8; SBN 15).[23]

In any event, a proposition such as *Hume is Hume* is trifling because composed of only one idea.

However, an affirmation of identity implies that the proposition has a subject and a predicate, according to Hume. Using a sentence of the form appropriate for expressing the proposition, *an object is the same with itself*, legitimately raises the expectation that the proposition expressed will not be trifling. It will be one expressing an identity we can be unsure about. Standardly (by Hume's lights), propositions about identity are propositions that differing things are the same thing. Standardly, a statement of identity says of something that something differing from it is, nonetheless, it. The something must be represented by one idea and the something differing must be represented by another idea. So a proposition about a standard case of identity must apparently consist of two ideas in order that something be added to the conception of the subject.

Hume concludes, "One single object conveys the idea of unity, not that of identity." A single object, perceived as such, causes only a single idea.[24]

However two or more objects perceived as such cannot convey the idea of identity either.

> On the other hand, a multiplicity of objects can never convey this idea, however resembling they may be suppos'd. The mind always pronounces the one not to be the other, and considers them as forming two, three, or any determinate number of objects, whose existences are entirely distinct and independent.
>
> (T 1.4.2.27; SBN 200)

Two or more ideas would represent there as being two or more things. The mind would have "an idea" of number. It is obvious that objects represented as two or more distinct objects are not represented as identical. So the mind cannot form an identity proposition consisting of two or more ideas.

> Since then both number and unity are incompatible with the relation of identity, it must lie in something that is neither of them. But to tell the truth, at first sight this seems utterly impossible. Betwixt unity and number there can be no medium; no more than betwixt existence and non-existence. After one object is suppos'd to exist, we must either suppose another also to exist; in which case we have the idea of

number: Or we must suppose it not to exist; in which case the first object remains at unity.

(T 1.4.2.28; SBN 200)

Thus, the mind apparently cannot form a nontrifling proposition that represents there as being a standard case of identity. It can form the proposition neither with one idea nor with more than one. One idea is too few to represent there as being the differing between the things said to be identical. Two ideas are too many to represent there as being the oneness of things said to be identical. But there is no middle way. Once the mind has one idea, there is either an additional one or there is not. If the mind represents there as being anything at all, it must represent there as being one thing or more than one. So the mind cannot conceive of identity as something it can be unsure about. Using one idea precludes the conceivability of the absence of the identity; using two ideas precludes the conceivability of its presence.

The differences between Hume's difficulty and Frege's puzzle are now even clearer. Hume's concern is how it is possible to represent there as being a standard case of identity—one in which there is some difference between the things taken to be identical. Frege's concern is how statements of identity can be both true and informative. The problem for Hume occurs at the level of intentional objects: what the propositions represent there as being. The problem for Frege occurs at the level of semantic content: what the semantic content is of referring expressions, which for both referent and sense is the level of intended (as opposed to intentional) objects. Despite the superficial close resemblance in presentation, these are two very different problems.

Failure to distinguish intentional object from intended object makes the difference almost impossible to see. When thinking in terms of intended objects, one sees Hume's move from distinct ideas to distinct objects simply as a mistake. And it would be a mistake. Obviously, distinct ideas could have the same intended object. Obviously, an identity proposition composed of distinct ideas could still be true. Obviously, Hume would be wrong if he were contending, as Bennett says he is, that "'x is y' is always false."[25] Hume's move, if this were it, would be not just a mistake, but a blunder. No theorist of ideas would need the technical notion of Fregean sense to see the blunder, contrary to Bennett's suggestion in trying to remove its sting. But moving from distinct ideas to distinct intended objects is not Hume's move. Hume is concerned with the number of intentional objects—how many objects we represent there as being. With distinct ideas, we represent there as being distinct objects. We do so even if there are not distinct intended objects. So we cannot represent there as being an identity with a proposition composed of two ideas. On this reading, Hume does not blunder.

Furthermore, Hume would likely have been aware of the semantic approach to such problems. Likely, such an approach was in the air. Close to Hume's difficulty is a problem Berkeley considers in the first draft of the

Principles. Berkeley is out to prove that belief in abstract general ideas is the result of the assumption that every word stands for an idea. He tries to come up with cases in which not every word in a statement can stand for a *particular* idea, in order to explain how the assumption motivates belief in *abstract* ideas. Then he goes on to show how to avoid belief in abstract ideas by rejecting the assumption. He picks the sentence "Melampus is an animal", where Melampus is "some one particular dog." He says that if "animal" "be made stand for another different [idea] from that is marked by the name Melampus, the proposition is false and includes a contradiction. And if it be made signify the very same individual that Melampus doth, it is a tautology." Berkeley's solution involves semantic ascent. He rejects the move that "animal" must therefore stand for an abstract idea. He says rather that the word "animal" does not stand for an idea at all. "All that I intend to signify thereby being only this, that the particular thing I call Melampus has a right to be called by the name animal."[26] Berkeley summarizes such a view in his Notebook A: "Homo est Homo etc comes at last to Petrus est Petrus etc. Now if these identical propositions are sought after in the mind they will not be found. there are no identical mental propositions tis all about sounds & terms" (*Works*, 728). Berkeley has extended a move made by Locke. Locke suggested that some trifling propositions, such as "*Parsimony is Frugality*" or "*Gold is a Metal*," are

> barely about the signification of Sounds. For since no abstract *Idea* can be the same with any other but it self, when its abstract Name is affirmed of any other Term, it can signify no more but this, that it may, or ought to be called by that Name; or that these two Names signify the same *Idea*.
>
> (*Essay*, 4.8.12)

Whether or not Berkeley's discussion would have been known to Hume, he would have known the Locke. So it is unlikely that if a semantic problem were really what Hume had in mind, he would have been too confused to propose a semantic solution.

The best way to bring out the differences in Hume's, Frege's, and Berkeley's approaches is to consider how one would go about conceiving the two possibilities concerning the identity—the possibilities whether it holds or does not. For Hume, if the proposition is not trifling, then we must be able to conceive the subject idea's intentional object and the predicate idea's intentional object—i.e., the things we represent there as being—as identical and yet also be able to conceive of them (the same them) as distinct. Hume's concern is with the intentional objects of the ideas that compose the propositions. We keep the intentional objects fixed between the conceived possibilities. Frege, on the other hand, would only be concerned with the possibilities concerning the statement of identity. We must be able to conceive the statement to be true and be able to conceive it to be false. If we conceive it true then we conceive there to be just one thing that both the senses of the referring expressions pick out. If we conceive it false, we conceive one sense of one referring expression to pick out one thing and the other sense of the other expression to pick out a second thing. We keep the statement and the senses of its terms fixed between the conceived possibilities. With Berkeley, if the identity holds then we have the right to call whatever particular thing he calls "Melampus" by the name "animal." If the identity does not hold, then we lack the right to call whatever he calls "Melampus" by the name "animal." What are held fixed between the possibilities are the definite description "the particular thing I call Melampus," and the name "animal."

So, Hume's is a different difficulty requiring a different sort of solution. What seems to be needed is an idea of a middle way between being a single thing and being more than one thing. Somehow we must have an idea of things that are more than just one single thing, but not quite two distinct things. Given that the number of ideas represents there as being that number of things, an idea must itself be "a medium betwixt unity and number," in order to be an idea of identity. But there is no such medium. So how can we have an idea of identity?

Removing the difficulty

Hume begins his answer with the confident phrase, "To remove this difficulty." This is overconfidence, I will argue, unless by "remove" he means displace. In any event the answer crucially involves the idea of time and a fiction.

> To remove this difficulty, let us have recourse to the idea of time or duration. I have already observ'd, that time, in a strict sense, implies succession, and that when we apply its idea to any unchangeable object, 'tis only by a fiction of the imagination, by which the unchangeable object is suppos'd to participate of the changes of the co-existent objects, and in particular of that of our perceptions.
>
> (T 1.4.2.29; SBN 200–1)

Strictly speaking, anything that endures is a succession. Thus, supposing that a steadfast object—a nonsuccession—endures is false. However, we have come naturally to make, and firmly to believe, this supposition about steadfast objects during the period before we ever paid close attention to their steadfastness. That is, when we think of a steadfast object, we have come to do so not with a steadfast idea, except right at first, but with a substituted succession of ideas—likely a succession of exactly resembling ideas, though Hume doesn't say so: "This fiction of the imagination almost universally takes place; and 'tis by means of it, that a single object, plac'd before us, and survey'd for any time without our discovering in it any interruption or variation, is able to give us a notion of identity" (T 1.4.2.29; SBN 201).

The fiction is the conceptual preparation needed for getting the idea of identity. Now all that is needed is the following experience: We pay attention to a steadfast object while some other background succession trips on—one we are not oblivious to. Perhaps we study a cup while the clock ticks. This experience causes the idea of identity. It doesn't matter whether the object really is steadfast. All we would need is an object that we represent as being steadfast because we don't "discover" that it is otherwise. We could get the idea from the successive images in a movie about a steadfast object. The important thing is that for the first time in our developing mental life we represent there as being a steadfast object and attend for a time to its steadfastness, while still fully convinced of its duration, and yet without slipping unawares into simply thinking of it as enduring, as we habitually did before.

When thinking back over the experience, we find that we represent the object as being a single thing or as being more than one thing, depending on how we think of the moments of time.

> For when we consider any two points of this time, we may place them in different lights: We may either survey them at the very same instant; in which case they give us the idea of number, both by themselves and by the object; which must be multiply'd, in order to be conceiv'd at once, as existent in these two different points of time: Or on the other hand, we may trace the succession of time by a like succession of ideas, and conceiving first one moment, along with the object then existent, imagine afterwards a change in the time without any *variation* or *interruption* in the object; in which case it gives us the idea of unity.[27]
>
> (T 1.4.2.29; SBN 201)

Interpreting this passage is a complicated and delicate business that draws on Hume's conceptions of duration, of steadfast objects, the relevant "fiction of the imagination," and of representing there as being number and unity.

First, we think simultaneously of two moments of the time while the steadfast object was surveyed. We think, say, of the moment at which one tick of the clock occurred and the moment at which a later one occurred. We think of the steadfast object as existent in these moments the way something with duration is: First it coexisted just with the one to the exclusion of any other, then it coexisted with the next to the exclusion of any other.[28] To think of the steadfast object as occupying each moment to the exclusion of the other, it "must be multiply'd." That is, we must represent it as being two distinct objects each of which occupies just one of the moments. Presumably, we represent it as two objects by means of distinct, simultaneous, exactly resembling ideas of it. As a result, we have an idea of number, both from thinking of the many moments and from thinking of the object as having duration by existing at each of them.[29]

On the other hand, we can think of the moments of the clock ticks successively without "multiplying" the steadfast object. Instead we think of the steadfast object as existing along with the succession of time, that is, as coexistent with the ticks' moments in the way characteristic of steadfast objects. So we conceive of the different moments by means of distinct ideas of the changing "co-existent objects" (the ticks), whereas the idea of the steadfast object will just be a single continued idea (cf. T 1.4.2.33; SBN 203). This point would not be clear without detailed consideration of Hume's conception of steadfast object and of representing there as being number. In isolation, the passage at T 1.4.2.29; SBN 201 is ambiguous. Another apparent interpretation is that the "like succession of ideas" is a gapless series of exactly resembling ideas of the steadfast object, and important commentators such as Stroud have read it this way. However, such a series would still "multiply" the object. Instead, we represent it as being single and steadfast despite the succession of coexistent objects. As a result, the idea of the steadfast object as steadfast gives us an idea of unity.

Thus, on the second view, we represent the object as being steadfast, that is, as being one single thing coexisting with a succession. Having studied it, we no longer simply confound it with something having duration. The pressure to regard the object as having duration cannot be denied, however, and manifests itself in the first view. In that view we think of the object not merely as coexisting with the distinct moments but as occupying each of them, that is, as being something with duration, that is, as being many distinct things.

"Here then is an idea, which is a medium betwixt unity and number; or more properly speaking, is either of them, according to the view, in which we take it: And this idea we call that of identity" (T 1.4.2.29; SBN 201). Thinking of the steadfast object as having duration, that is, as occupying distinct moments, is representing it as being many distinct things and not steadfast. Thinking of it as steadfast, that is as invariable and uninterrupted, is representing it as being a single thing and not enduring. These ways it is represented as being are contradictory, so the fiction of an object represented as being both enduring and steadfast requires switching back and forth between viewing the object as many and viewing it as one. Since there cannot be an idea literally of a medium between singleness and plurality, this mongrel, view-dependent idea serves well enough. The switching of views masks the inconsistency.

"We cannot, in any propriety of speech, say, that an object is the same with itself, unless we mean, that the object existent at one time is the same with itself existent at another" (T 1.4.2.29; SBN 201). Only this view-dependent idea involving time captures well enough the dueling aspects an idea of identity must have. So when attributing identity, we must represent there as being distinct objects existing at distinct times, on one view, which are a single continued object coexisting with those times, on the other view. One must be able to take both views, make use of their differences, and switch between them, so that each can supplement the other, in order to apply the idea of identity properly.

"By this means we make a difference, betwixt the idea meant by the word, *object*, and that meant by *itself*, without going the length of number, and at the same time without restraining ourselves to a strict and absolute unity" (T 1.4.2.29; SBN 201). The words "object" and "itself" signify two different ideas on one view and one idea on the other view. On the first view, the word "object" signifies an idea of the object occupying one moment, and the word "itself" signifies an idea of the "multiply'd" object occupying a distinct moment. On the second view, the words "object" and "itself" both signify the idea kept steadfast. So the meanings are not simply two nor simply one. The so-called "medium" is that the meanings are two on one view and one on the other. Thus, the proposition *an object is the same with itself* is composed of two ideas on one view and one idea on the other. Consequently, what the mind represents there as being are two on one view and is one on the other.[30]

> Thus the principle of individuation is nothing but the *invariableness* and *uninterruptedness* of any object, thro' a suppos'd variation of time, by which the mind can trace it in the different periods of its existence, without any break of the view, and without being oblig'd to form the idea of multiplicity or number.
>
> (T 1.4.2.30; SBN 201)

Here Hume gives the characteristics we must experience in an object in order to acquire the idea of identity: invariableness, uninterruptedness, and existence at distinct times. These collectively are the "principle of individuation" in that they are the causes of the idea of identity. Hume tends to use "principle" to mean cause.[31] The third of these characteristics involves fiction, which is why Hume speaks of "a *suppos'd* [my emphasis] variation in time." The experience itself is of an invariable and uninterrupted, that is, steadfast, object. Such an object is what one represents there as being because one experiences no noticeable alteration and no noticeable succession and tends to use a single continued idea when possible (T 1.4.2.29, 1.4.2.33; SBN 201, 203).

Hume says this experience does not oblige us to form the idea of multiplicity or number. Still, because of the supposed variation in time, the supposed duration, there is the option of viewing the object as multiple. That option is what prevents the identity of the object from degenerating into oneness, and so enables us not only to conceive that the identity holds, but also to be able to conceive that it does not. Nonetheless, the view of the object as unitary is the dominant view; it is stronger since it is directly copied from experience.

This is Hume's account. I end by considering three objections—two that Hume can answer and one, I think, he cannot.

Three objections

It appears to some commentators that Hume begs the question in his account of the acquisition of the idea of identity through time. Part of the problem, as noted above, is that Hume seems to talk as if the second view of the enduring, steadfast object—which is just part of the idea of identity—is the whole of the idea. So it seems that the whole is presupposed in an account of the whole. But there is more to the objection as well. It is based on the assumption that perceptions are either exceedingly brief, or are composites of exceedingly brief perceptions. Green, Stroud, and Bennett share this assumption, the "Brevity Assumption."[32]

I have shown that perceptions may well be steadfast. We need to distinguish temporally lengthy single perceptions that do not endure, from those successions of perceptions that do endure. Still the charge of circularity against Hume has been an important one, so it is worthwhile understanding it and where it goes wrong.

Green runs together Hume's account of the acquisition of the idea of identity, with his account of the application of that idea to a succession of related perceptions.[33] This fully intended confusion can be explained by his understanding of Hume's claim that "impressions are … perishing existences" (T 1.4.2.15; SBN 194). Green assumes that for Hume no perception (impression or idea) is a unity if it coexists with a succession. He assumes, rather, that any such perception is itself really a succession of perceptions. His assumed interpretation makes Hume seem to him to beg the question as follows: To acquire the idea of identity through time is to acquire the idea of something which is unitary and exists at a succession of times. Hume says we get the idea from an impression of a steadfast object. But any such impression which might give this idea is itself really a succession of related impressions. A succession of related impressions can convey the idea of identity only by being confused with a unitary impression which exists at a succession of times. Such confusion would be possible only for someone who already had the idea of something unitary which exists at a succession of times. But this is already to have the idea of something with identity. So the idea of identity can be acquired only if it is already had.[34]

There are gaps in this argument, but its main point is clear: If we assume that perceptions are exceedingly brief, then Hume begs the question.

The role of the Brevity Assumption in Stroud's argument is not immediately apparent, but I think it is crucial. Stroud begins by noting that on Hume's definition, identity is the "*invariableness* and *uninterruptedness* of an object, thro' a supposed variation of time" (T 1.4.2.30; SBN 201). However, on Hume's account of acquiring the concept of identity, the acquisition involves imagining "a change in time without any *variation* or *interruption* in the object" in the second light in which the moments are surveyed (T 1.4.2.29; SBN 201). But that just is, by definition, to imagine the object's identity through the change in time. So acquiring the concept of identity requires an act of imagination involving the concept of identity. So, Stroud concludes, on Hume's account the concept of identity can be acquired only if it is had.[35]

Stroud presupposes, incorrectly, that in the second view is an idea of the single object with duration. Rather, it is only an idea of a single, steadfast object coexisting with a succession. To think of the object from the second view requires just the sort of idea that Hume says is used to think of a steadfast object: namely, a single, continued idea (T 1.4.2.33; SBN 203). I conjecture that Stroud goes wrong because of the Brevity Assumption. If perceptions were brief, then the only way to imagine a steadfast object would be by means of a succession of perceptions. But this would be no different than imagining the steadfast object to be also having duration. For, if the idea is a succession then it can serve as an idea of duration (T 1.2.3.10–11; SBN 36–7). Thus, Stroud cannot distinguish the second view of the steadfast object from the idea of identity itself, so cannot give a subtle enough rendering of Hume's account to prevent apparent question-begging.

Second objection: Hume says that the fiction of a steadfast object with duration is the only way to address the difficulty with the idea of identity. However, it might seem that making a distinction of reason would be an additional way for Hume to "make a difference, betwixt the idea meant by the word, *object*, and that meant by *itself*."[36] Not so. A distinction of reason is merely a conceptual distinction. Making a distinction of reason is simply a matter of distinguishing distinct resemblance classes that some one thing belongs to. This is not a way of considering the possibility that the thing is really two distinct things. Representing something as two, requires two ideas. The "ideas" used to think of things distinguished by a distinction of reason "can neither be separable nor distinguishable." That is, there is only one idea, considered in the light of different resemblances (T 1.1.7.17–18; SBN 24–5). Thus, making a distinction of reason cannot help with representing the possibility of there being a numerical distinction, which is what is at issue for Hume.

Third objection: Hume's account explicitly involves a fiction. Thus, he has not given an account that strictly speaking resolves the difficulty. He has only explained how we generate an idea that seems to resolve it. Recall that we must be able to think of some things while leaving it open whether or not they are identical with each other. So we must be able to represent there as being things that are the same things as themselves whether we consider them as one and the same with each other, or as two things distinct from each other. It is them we view either way, and we must be able to be aware of that. Hume has explained how we come up with an idea that involves switching between the view of something as one and the view of some things as many, but he has not explained how we represent the something viewed as one and the things viewed as many as the same thing(s). That, however, was the original problem.

It does no good to point out that the various ideas of the steadfast object in the two views have the same intended object. That is true, but the problem is not with what there is that we represent, but with what we represent there as being. The problem is at the level of intentional objects.

The fact that the ideas of the "multiply'd" objects and the idea of the continued object all exactly resemble (except for temporal length) is not enough to represent them as the same thing(s). Without some further way of representing the identity, they would rather be representing the continued object as yet one additional thing.[37]

It might seem that Hume could appeal to his mechanism of confounding to explain how we represent the sameness of the one with the many. But confounding is what gave us the "idea of time without a changeable existence" (T 1.2.5.29; SBN 65), that is, the idea of something with duration being inadvertently substituted for an idea of a steadfast object, which was the prerequisite "fiction of the imagination" (T 1.4.2.29; SBN 200–1). Acquiring the idea of identity required resisting the confounding in order to appreciate the object's steadfastness. The basis of confounding is our *not* noticing the difference between an idea and its substitute idea. In contrast noticing the difference between the two views of the steadfast object is precisely what is required for acquiring the idea of identity.

That we need to notice the difference may well be why Hume has it that the first view we take of the idea of identity, involves considering different moments *simultaneously*. Strictly speaking, we could consider the moments successively and think of the steadfast object as an accompanying succession. But this would be too easy to confuse with the idea of identity in the second view, in which we merely continue a single idea of a steadfast object along with the successive ideas of the individual moments. To distinguish easily the two views, we need the first to be such that "the progress of the thought is broke" (T 1.4.3.4; SBN 220). Thus, Hume seems to describe the idea in such a way to prevent confounding, not to rely on it.

So Hume's attempt to evade his difficulty with identity is faced with the same difficulty. He can't explain how we can represent there as being things that are perhaps one single thing and perhaps two distinct things. The difficulty cannot be finessed without being answered.

We need ideas that are actually a "medium betwixt unity and number" in order genuinely to represent there as being such a medium, in order fictitiously to represent there as being such a medium. Yet, as Hume says, "Betwixt unity and number there can be no medium."

5 Representing personal identity

Famously, Hume finds a contradiction in his theory of the intellectual world, as he explains in an enigmatic passage in the *Treatise*'s Appendix. He says he finds himself in a labyrinth concerning personal identity. The passage has proved wonderfully elastic, accommodating a variety of interpretations, though none comfortably. I will try for a somewhat better fit by giving a simple argument that stays very close to Hume's own words in the passage and that adds only an assumption he endorses elsewhere. I focus on Hume's account of consciousness. No competing interpretation has had this focus despite the fact that the "promising aspect" of Hume's account is that "personal identity *arises* from consciousness" and that his "hopes vanish" when he comes "to explain the principles, that unite our successive perceptions in our thought or consciousness" (T App. 20; SBN 635–6). The key assumption for Hume, I argue, is that consciousness's ideas are unerring. The Appendix labyrinth is how to make sense of an identity between the distinct past perceptions, in order to explain how consciousness's unerring ideas can represent them as identical. If I am right then, intriguingly, it looks as if the Appendix contradiction can be resolved only if the previous difficulty with identity can be.

Interpreting the Appendix passage

In "Of Personal Identity," Hume explains that the self is really nothing but many distinct perceptions. Because of their close relations we take them to be identical. By this means we arrive at the idea of the self as one, identical thing. The result is a fiction: "that identity is nothing really belonging to these different perceptions, and uniting them together; but is merely a quality, which we attribute to them, because of the union of their ideas in the imagination, when we reflect on them" (T 1.4.6.16; SBN 260).

In other words, we reflect on "the train of past perceptions" (T App.20; SBN 635). The ideas reflecting them have a "union ... in the imagination" (T 1.4.6.16; SBN 260). So, we falsely attribute identity to the perceptions reflected on. In the Appendix passage, however, Hume finds himself unable to explain our attribution of identity to these many perceptions reflected on: "when I proceed to explain the principle of connexion, which binds them together, and makes us attribute to them a real simplicity and identity, I am sensible, that my account is very defective" (T App.20; SBN 635).[1] He confesses, "I neither know how to correct my former opinions, nor how to render them consistent," though he has hopes that eventually someone may find a way to "reconcile those contradictions" (T App.10, App.21; SBN 633, 636).

For obscure reasons, Hume thinks that to correct the defect in his account, the perceptions reflected on must either "inhere in something simple and individual" or be such that the mind can "perceive some real connexion among them." The interpretive challenge is to make the reasons clear, that is, to explain what the contradiction is, how one of these proposals would resolve it, and why he can't accept either proposal. As I will argue, the contradiction is that an explanation of how we represent the distinct past perceptions as identical, requires that they somehow be identical. The two proposed corrections are two ways to try to explain how distinct things can nonetheless be identical. Hume cannot accept either because he has already, for weighty reasons, rejected both. The hidden key to understanding the contradictions is Hume's implicit assumption that consciousness's ideas, unlike various other ideas of perceptions, are accurate.[2]

In giving my interpretation, I will call the perceptions reflected on, the "past perceptions." I will call the ideas reflecting them "consciousness's ideas," since "consciousness is nothing but a reflected thought or perception" (T App.20; SBN 635).[3] That is, consciousness's ideas are perceptions of perceptions, in accordance with Locke's statement, used by Johnson in his definition, "Consciousness is the perception of what passes in a Man's own mind" (*Essay* 2.1.19).[4] I will put the interpretation as a set of sentences consisting of (1) the claim whose explanation is defective, (2) the assumption that I propose is implicit in the passage, and (3) and (4) the two principles such that he says holding both makes his explanation defective. The inconsistency is made explicit by deriving (5) from (2), (3), (4), then deriving (6) from (1), (2), (5).

1. Consciousness's ideas represent the distinct past perceptions as being identical with each other.
2. Consciousness's ideas represent the past perceptions as they are.
3. Distinct perceptions are distinct existences.
4. The mind never perceives an identity (a "real connexion") between distinct existences.
5. So consciousness's ideas represent the distinct past perceptions as being not identical with each other.
6. So the distinct past perceptions are identical with each other and are not identical with each other.

It may well seem that (6) follows directly from (1) and (2). However, with (3) and (4), Hume objects to theories that, if correct, would make (6) either not really follow or not really be a contradiction.[5]

The background to sentence (1) occurs in "Of Personal Identity." Hume has briefly argued that there is no simplicity in the mind at one time, nor identity between different times. "They are the successive perceptions only, that constitute the mind" (T 1.4.6.4; SBN 253). The bulk of T 1.4.6 is devoted to answering the question, "What then gives us so great a propension to ascribe an identity to these successive perceptions, and to suppose ourselves possest of an invariable and uninterrupted existence thro' the whole course of our lives?" (T 1.4.6.5; SBN 253). Hume's answer is that "the train of past perceptions" (App.20; SBN 635) is a case of "several different objects existing in succession, and connected together by a close relation" (T 1.4.6.6; SBN 253), or in this case, two close relations—resemblance and cause and effect (T 1.4.6.17–19; SBN 260–1). These close relations give them a felt connection, so much so that they make the succession feel to the mind as if it were contemplating an uninterrupted and invariable object. This resemblance in feeling "makes us substitute the notion of identity, instead of that of related objects" (T 1.4.6.6; SBN 254). This is one of the promised "many instances of this tendency of relation to make us ascribe an identity to different objects" (T 1.4.2.35; SBN 204).[6] We try to correct this tendency, but it is very strong. "Our last resource is to yield to it, and boldly assert that these different related objects are in effect the same, however interrupted and variable" (T 1.4.6.6.; SBN 254). It is this representing the distinct past perceptions as identical that Hume means to be explaining.[7]

In doing so, we do not regard any of our perceptions as, e.g., "a hat, or shoe, or stone" as we tend to regard our impressions of sense in other circumstances (T 1.4.2.31; SBN 202). Rather as Hume says, "In thinking of our past thoughts we not only delineate out the objects, of which we were thinking, but also conceive the action of the mind in the meditation, that certain *je-ne-sçai-quoi*, of which 'tis impossible to give any definition or description, but which every one sufficiently understands" (T 1.3.8.16; SBN 106).

Presumably we consider the "action of the mind" in recollecting any past perception as such, not just ideas. In doing so, we consider them "as real perceptions in the mind, of which we are intimately conscious" (T 1.3.8.15; SBN 106).[8] So we take distinct past perceptions *considered as perceptions in the mind*, to be identical, when arriving at the idea of self as one, identical thing.

Again, he later finds his account defective when proceeding "to explain the principle of connexion, which binds them together, and makes us attribute to them a real simplicity and identity" (T App.20; SBN 635). Originally, in "Of Personal Identity," the combination of resemblance and causation plays this binding role. "Now the only qualities, which can give ideas an union in the imagination, are these three relations [resemblance, contiguity, and causation] above mentioned," and he later says contiguity is not relevant here (T

1.4.6.16–17; SBN 260). So the question is, why cannot resemblance and causation properly play the role of the principle of connexion?

Originally the "union" he was speaking of was the felt connection. So it might appear that Hume later is finding some problem with causation and resemblance giving the felt connection. Given the ambiguities of the Appendix passage, one might think Hume means that there is not sufficient causation or resemblance between the past perceptions to yield any felt connection in the first place.[9] However, if Hume were raising this problem, he would have talked more about causation and resemblance in the passage and would not have characterized his brief summary of the felt connection between the past perceptions as having a "promising aspect." His "hopes vanish" only when shifting from the level of past perceptions to the level of consciousness's ideas of them. So were there any problem with felt connection, it would likely be at this higher level of reflecting perceptions. This level of "thought or consciousness" is a natural one to consider. After all, a felt connection between reflected perceptions results from a felt connection between the reflecting ideas, which consists in the fact that the reflecting ideas "naturally introduce each other." In other words, the feeling of connection between the reflected perceptions is "spread" on them and is really "the determination of the mind" to pass from one reflecting idea to the next (cf. T 1.3.14.25; SBN 167). Is Hume finding a problem, then, with resemblance or causation giving a felt connection between the reflecting perceptions? That might have been plausible had he never considered this level before. But he had considered it, at least for causation. He specifically examined what the "determination of the mind" was and found it just likewise to be a "new determination" at a higher level (T 1.3.14.29; SBN 169). If Hume thought there were problems with higher-level determinations of the mind, he would have found it here or else would have found it on review of his causation section, not his personal identity section.

No, in the subsequent Appendix passage, he never questions the ability of resemblance and causation to make the train of past perceptions feel so connected that it feels like having a steadfast perception. Rather, he is finding that his account of felt connection is inadequate to explain how we represent the past reflected perceptions as identical. Some additional quality, "which can give ideas an union in the imagination" seems necessary.[10] To see what it is, we need to look at what he focuses on in the Appendix passage when trying to find where he went wrong. He does not explore the details of his account of causation and resemblance and felt connection. Instead, he focuses in great detail on his reasons for regarding the past perceptions, as well as all distinct perceptions, as distinct existences, and his reason for thinking the self is just a composition of perceptions. These are the arguments that induced him "to deny the strict and proper identity and simplicity of a self or thinking being" (T App.10; SBN 633). It is here that he thinks he might have made a misstep. For he plans to "propose the arguments on both sides," so will next propose those for *affirming* the identity of the self. Thus the principle of connexion that he is worried that he needs is identity itself.

But why identity? Sentence (2) captures my proposed answer. Hume is taking for granted that consciousness's ideas accurately reflect the perceptions they are of. He has already stated and reiterated that he expected there to be no contradictions in his theory of the mind, unlike "in every system concerning external objects" (T 1.4.5.1; SBN 232, see T App.10; SBN 633). He explains his expectation in "Of Benevolence and Anger."

> The essence and composition of external bodies are so obscure, that we must necessarily in our reasonings, or rather conjectures concerning them, involve ourselves in absurdities and contradictions. But as the perceptions of the mind are perfectly known, and I have us'd all imaginable caution in forming conclusions, concerning them, I have always hop'd to keep clear of those contradictions, which have attended every other system.
>
> (T 2.2.6.2; SBN 366)

The root of the hope is his conviction that the perceptions of the mind, considered as such, are "perfectly known." I take this to mean that they are reflected accurately in consciousness. Certainly, once the imagination takes over, any number of Humean fictions about our perceptions can be generated. These fictions get us into trouble in our reasonings concerning external objects. But consciousness itself is our access to our own minds, and it is accurate. "For since all actions and sensations of the mind are known to us by consciousness, they must necessarily appear in every particular what they are, and be what they appear" (T 1.4.2.7; SBN 190). Even in the *Enquiry*, he is still saying, "But consciousness never deceives" (EHU 7.13; SBN 66).

Hume in some places *seems* to say that we can go wrong about our perceptions or the operations of the mind. He seems to say, for instance, that we confound two sentiments "from interest and morals" (T 3.1.2.4; SBN 472). But in such a case we are not merely conscious of the sentiments themselves; rather we confound them when our thoughts are about an enemy and we fail to distinguish his opposition to our interest from villainy. In another place, he says, "there are certain calm desires and tendencies" that "are very readily taken for the determinations of reason" because "their sensations are not evidently different" (T 2.3.3.8; SBN 417). But this happens when later classifying these actions of the mind in terms of their similarity in provoking little or no emotion, and thus neglecting the difference between the actions themselves. In an additional place, he says that it is even harder than in the physical world to discern causes in the mind,

> where there is a much greater complication of circumstances, and where those views and sentiments, which are essential to any action of the mind, are so implicit and obscure, that they often escape our strictest

> attention, and are not only unaccountable in their causes, but even unknown in their existence.
>
> (T 1.3.15.11; SBN 175; see also EHU 1.13; SBN 13)

Here, however, he likely means "unknown" to the theorist, not the subject. In most of these places Hume is acknowledging difficulties with theorizing about perceptions—classifying them and explaining their causes and effects—not with the mere consciousness of them.[11] It is about such cases of theorizing that he observes, "'tis very difficult to talk of the operations of the mind with perfect propriety and exactness; because common language has seldom made any very nice distinctions among them, but has generally call'd by the same term all such as nearly resemble each other" (T 1.3.8.15; SBN 105). So, in none of these places does Hume say anything incompatible with his assumption that consciousness never deceives.

Further reason that Hume is taking accurate reflection for granted is that the ideas of consciousness he is presently concerned with in "Of Personal Identity" and the Appendix passage are ideas of memory: "As memory alone acquaints us with the continuance and extent of this succession of perceptions, 'tis to be consider'd, upon that account chiefly, as the source of personal identity" (T 1.4.6.20; SBN 261).

Hume is, of course, evoking Locke's discussion of consciousness of past perceptions, i.e., memory, which is central to the *Essay*'s account of personal identity.[12] One of the hallmarks of memory is accurate reflection. "[O]ur confidence in the veracity of that faculty is the greatest imaginable" (T 1.3.13.19; SBN 153). Hume asks rhetorically, "For what is the memory but a faculty, by which we raise up the images of past perceptions?" and states in a subordinate clause, "an image necessarily resembles its object" (T 1.4.6.18; SBN 260). "'Tis evident, that the memory preserves the original form, in which its objects were presented, and that wherever we depart from it in recollecting any thing, it proceeds from some defect or imperfection in that faculty" (T 1.1.3.3; SBN 9).

Hume appeals to no defect or imperfection in the faculty of memory in the Appendix passage, so, I suggest, he is taking the accuracy of the reflection of the past perceptions for granted. He retains this view into the *Enquiry*. "When we reflect on our past sentiments and affections, our thought is a faithful mirror, and copies its objects truly" (EHU 2.2; SBN 17–18).[13]

Nor should we think that Hume *could* appeal to a defect in the memory of past perceptions in order to escape the labyrinth I say he is in.[14] The intimacy of our acquaintance with the perceptions composing the self is central to his Book 2 project of explaining our passions. For instance, when discussing the object of pride and humility, he says, "This object is self, or that succession of related ideas and impressions, of which we have an intimate memory and consciousness" (T 2.1.2.2; SBN 277). The intimacy of the acquaintance gives the idea of self an extraordinarily high degree of vivacity.

"'Tis evident, that the idea, or rather impression of ourselves is always intimately present with us, and that our consciousness gives us so lively a conception of our own person, that 'tis not possible to imagine, that any thing can in this particular go beyond it" (T 2.1.11.4; SBN 317). Hume often appeals to this high degree of vivacity in explaining the course of various passions (T 2.1.11.4, 2.2.2.15, 2.2.2.16, 2.2.4.7, 2.3.7.1; SBN 320, 339, 340, 354, 427). It is crucial to his account of sympathy, which plays such an important role in explaining our approval of many natural as well as artificial virtues (T 3.3.1.10–11; SBN 577–8). Given this vivacity, it follows from his account of memory that memories of consciousness will tend to be accurate. Even more than the average memory, they will easily "be preserv'd by the mind steady and uniform for [a] considerable time," unlike the "faint and languid" ideas of the imagination whose lack of vivacity permits more variation (T 1.1.3.1; SBN 9).

Even if Hume allowed that our consciousness of past perceptions were more fallible, that could not help him out of the labyrinth. Whatever perceptions might be forgotten or features misremembered, the multiplicity of the past perceptions—their distinctness from each other—is too manifest to be forgotten. According to Hume's original "defective" account, when attributing identity we start out regarding "the related succession as variable or interrupted." Even in the course of yielding to the "mistake" of taking them to be identical, "we incessantly correct ourselves by reflection, and return to a more accurate method of thinking" (T 1.4.6.6; SBN 254). After all, "'Tis evident, that the identity, which we attribute to the human mind, however perfect we imagine it to be, is not able to run the several different perceptions into one, and make them lose their characters of distinction and difference, which are essential to them" (T 1.4.6.16; SBN 259). Consciousness need only accurately represent the *distinctness* of the past perceptions to get Hume in trouble, and that it surely does. There is no defect in memory that could explain consciousness making a mistake about their distinctness.

It might appear that the accurate reflection by consciousness of the past perceptions is compatible with beliefs spawned by the imagination that the perceptions are identical and made so by something unintelligible. In "Of Personal Identity," he speaks of substituting an idea of identity for an idea of the succession of related perceptions. The accurate reflection comes first, immediately followed by the substitution of an arbitrary idea of a steadfast object.[15] By means of the substitute idea we "feign some new and unintelligible principle, that connects the objects together, and prevents their interruption or variation." In the case at hand, the "objects" are the past perceptions. The idea of the unintelligible principle is "the notion of a *soul*, and *self*, and *substance*" (T 1.4.6.6; SBN 254). Swept along by the analogy with external objects such as "any mass of matter," a planet, a church, a ship, a river, "vegetables and animal bodies," Hume tries to give this same sort of account of our attribution of identity to "the mind of man" (T

1.4.6.15; SBN 259). He is only encouraged in this analogy by the third-person perspective that he uses to emphasize the difference between reflecting ideas and ideas reflected on, and from which he talks of "looking into the breast of another" (T 1.4.6.16; 1.4.6.18; SBN 259, 260).[16]

However, in giving this account, Hume overlooks the fact that he is trying to explain a mistake, not about any old succession of external objects, nor a succession within another person, but about a succession of things we are intimately conscious of—our past perceptions. Further the mistake is compounded by positing, not something "unknown and invisible" such as a pure substance (T 1.4.3.4; SBN 220), but rather something we are likewise supposed to be intimately conscious of. Hume is trying to explain the fictitious belief that "we are every moment intimately conscious of what we call our SELF; that we feel its existence and its continuance in existence; and are certain beyond the evidence of a demonstration, both of its perfect identity and simplicity" (T 1.4.6.1; SBN 251). Consciousness is no enemy of mistaken belief in the existence of "unintelligible principles" of identity in external objects or other minds. But the mistake that we are conscious of such a "principle" within ourselves would be easily exposed by the fact that we are not (T App.15; SBN 634). To suppose otherwise would be to suppose "that even where we are most intimately conscious, we might be mistaken"—something Hume regards as impossible (T 1.4.2.7; SBN 190). The attribution of identity to the distinct perceptions cannot be the result of mistakes in consciousness, nor mistakes unnoticed by consciousness.

Not only does Hume assume accurate reflection in consciousness, he is prone to leave the assumption implicit. Consider the famous passage when he reflects on himself. He perceives perceptions and only perceptions, and therefore unhesitatingly "venture[s] to affirm of the rest of mankind, that they are nothing but a bundle or collection of different perceptions" (T 1.4.6.4; SBN 252). He reiterates this move in the Appendix: "When I turn my reflexion on *myself*, I never can perceive this *self* without some one or more perceptions; nor can I ever perceive any thing but the perceptions. 'Tis the composition of these, therefore, which forms the self" (T App.15; SBN 634). The inference implicitly assumes that the objects of consciousness, including the self, are just as they appear to consciousness. I suggest that the assumption is implicit in the Appendix problem as well.

In the Appendix, Hume discovers his oversight. He now fully appreciates that it is objects of consciousness with which he is concerned. Thus, given the implicit (2), if consciousness's perceptions are to represent the past perceptions as identical, they must somehow reflect an identity in them already.

There needs to be a genuine identity in the reflected perceptions. But that contradicts one or the other of two dearly held principles: (3) Distinct perceptions are distinct existences, and (4) The mind never perceives an identity (a "real connexion") between distinct existences. He says these principles are "inconsistent," but since they are consistent with each other, he must mean they are jointly inconsistent with his account insofar as it needs identity in the reflected perceptions. Giving up (3) or (4) would make Hume's account consistent, but he doesn't see how he can give up either. So he cannot figure out how "to explain the principles, that unite our successive perceptions in our thought or consciousness." In other words, he cannot explain the identity between the reflected perceptions needed for the reflecting ideas to represent them as being identical with each other.

Giving up (3) would be allowing that the distinct perceptions "inhere in something simple and individual," in other words are modes of a single substance, as was believed by "the Theologians." The distinct perceptions would not be "distinct existences," so to some extent they would be the same existence. There would be an underlying identity in the mind, exactly like the underlying identity in Spinoza's world-God. Hume, however, has detailed the "absurdities" of such a view (T 1.4.5.17–25; SBN 240–4). Giving up (4) would likewise be allowing some sort of identity between the distinct perceptions. The denial of (4) that the mind perceives a real connexion is, in more perspicuous words, a denial that any "connexions among distinct existences are ever discoverable by human understanding" (T App.20; SBN 635). So giving up (4) is allowing that there is a real connexion there to be discovered. The real connexion in (4) is the "real bond" inquired after at T 1.4.6.16; SBN 259 and specified as "identity" both just before and two sentences later.[17] The view that there could be a mysterious identity between distinct existences was held by Shaftsbury and criticized by Hume at T 1.4.6.6; SBN 254 as an "absurdity," a view that motivates people to justify it by believing in a principle that is "unintelligible" and "unknown and mysterious."[18] Thus, either solution would require things that are somehow distinct, to be nonetheless somehow identical as well.[19] Hume mentions them because they are extant views that would give him the identity he needs. Either way, consciousness's ideas could reflect the distinctness of the past perceptions while yet representing them as identical.

Nonetheless, it is strange for Hume to say that adopting either view would make there be "no difficulty in the case," as if they were genuine options. They seem to involve the same contradiction he is trying to resolve—the bald contradiction that some things are not identical and are identical. It must be that he senses that there is something right about these views, despite their flaws. Perhaps he is sensing that the mind, as we are conscious of it, exhibits an identity-in-difference that he can't yet make sense of and that the two views he considers are imperfectly trying to capture.[20]

If there is such an identity-in-difference, then the perceptions distinct from each other in one respect are identical with each other in another. To make sense of this, one would genuinely have to be able to represent there as being things that are the same things whether they are represented as identical or represented as distinct. In more Humean words, one would genuinely have to be able to represent there as being a "medium betwixt unity and number." It appears that at the heart of Hume's inability to escape the labyrinth concerning the attribution of personal identity is his inability to give a genuine resolution to the prior difficulty concerning identity.[21]

Other interpretations

Hume's Appendix worry is how to make sense of an identity between the past perceptions, in order to explain how consciousness's unerring ideas can represent them as identical.[22]

If I am right, then interpretations of the Appendix concerned with other relations bundling perceptions into composite selves are off the mark. These include those of Basson, Beauchamp, Flage, Garrett, Haugeland, Loeb, MacNabb, Patten, Pears, Robison, and Stroud.[23] They raise hard problems for Hume, but solutions to them would not address the Appendix problem.[24] Hume is not worried whether his account includes the right things in the bundled self, or how he can individuate the bundles; he is worried that the self cannot merely be a bundle of distinct perceptions.

Likewise interpretations concerned with non-Humean selves bundling perceptions into selves, are off the mark. These include those of Johnson, Kemp Smith, Passmore, Price, and Robison.[25] They assume that only a unitary, active self not composed of perceptions could smoothly transit the many associated perceptions and confound them with a single thing, and they assert that in the Appendix Hume is realizing this. But such interpretations leave unanswered the problem how consciousness can represent the distinct past perceptions as identical. Further, their assumption just begs the question against Hume's view that the self is many, passive perceptions. As several commentators have pointed out, for Hume, association, transit, and confounding are just a matter of which perceptions tend to accompany or follow which; no further explanation is possible or necessary.[26]

Refutations of most of the interpretations cited so far are found in Garrett, Fogelin, and Penelhum.[27] In Garrett and Penelhum are refutations of Kemp Smith's claim that the account of self in Book 1 is incompatible with the awareness of self required in the Book 2 accounts of sympathy and of the indirect passions.[28] It seems to me that were the Book 1 account of the fictional idea of the self successful, that idea, sufficiently enlivened, could play any role required by Book 2. This much can be said for Kemp Smith's position, however. It may be that the relevant awarenesses of the self in Book 2 require a lively idea of the self as unitary. If so, then Kemp Smith would be right that the Book 2 appeals to the idea of self are inconsistent with the Book 1 account of the self as many. He would be right, that is, if I am right that Hume cannot explain how consciousness's ideas could represent the many perceptions as something simple and identical.

The problem is with the fictitious attribution by consciousness of identity to the many perceptions. There is a third group of interpretations, most not specifically of the Appendix, that can be used to argue that the identity attributed is not fictitious, so there is no false attribution to explain.[29] This group includes Ashley and Stack, Biro, Kemp Smith, Laird, MacNabb, Penelhum, Perry, and Swain.[30] They roughly divide into those who think identity, not perfect but imperfect, is rightly attributed either to the distinct perceptions or to the bundle of them as a whole, and those who think identity *simpliciter* is rightly attributed to the bundle of successive perceptions as a whole. Penelhum (the most important commentator on Hume on personal identity) and Perry think Hume failed to see that a bundle of differing parts could be identical through time. Thus he needlessly thought that the attribution of identity was fictitious.

The first subgroup thinks perfect identity is the sort of identity characteristic of unchanging steadfast objects, while imperfect identity is the sort characteristic of successions. But this is either confused or the position of the second subgroup. If imperfect identity is a relation between distinct things it is not identity no matter what you call it.[31] If it is a relation a succession has to itself then it is just identity regardless of what it is being applied to. In neither case are there two kinds of identity, perfect and imperfect. Nor did Hume believe in two kinds of identity. For him, perfect identity is just identity. Imperfect identity is just identity naturally but falsely believed to hold between distinct objects because of the relations between them.[32] It is, so to speak, imaginary identity found in "instances of this tendency of relation to make us ascribe an *identity* to *different* objects" (T 1.4.2.35; SBN 204).

Now in getting the idea of the self, what are the "different objects" to which identity is falsely ascribed? They are the "successive perceptions" (T 1.4.6.5; SBN 253).[33] The second subgroup neglects the point that the relevant relata of the attributed identity are perceptions, not bundles of them. Perhaps Hume could have given what we would regard as identity conditions for bundles of perceptions, qua bundles. But his concern is to explain why, when introspecting many successive things, we take them to be identical with each other.[34]

It is true that Hume seems *also* to talk about the fictitious identity of bundles to themselves. He does seem also to think that having a succession of parts is contrary to something's identity through time (T 1.4.6.7–14; SBN 255–8). He thinks this because he thinks that, strictly speaking, nothing with parts exists. Only the parts do (T 1.2.2.3; SBN 30). So the talk of a succession lacking identity boils down, for Hume, to the claim that the members of the succession are distinct. So the identity fictitiously attributed is the supposed identity of each of these members with the others.

To appreciate this point about Hume, one must see him as also thinking on a metaphysical level below the fictions of the commonsensical level. Were he operating just at the commonsense level, he would be subject to the criticism of Penelhum. At the commonsense level, it would certainly be mistaken—as Penelhum charges—for Hume to deny that the same thing can alter and to deny that a succession can truthfully be denominated a single thing. But far from denying these at the commonsense level, Hume explains the causes of our thinking these ways at that level. It is at the metaphysical level that he denies them. At that deeper level, variation is "evidently contrary" to identity (T 1.4.3.2; SBN 219) and "That term unity is merely a fictitious denomination, which the mind may apply to any quantity of objects it collects together" (T 1.2.2.3; SBN 30). Without taking on Hume's likely reasons for these claims, Penelhum cannot just assume that there are bundles that are single changing things, and so cannot just assume that Hume is in a muddle when denying it.[35]

A fourth group of interpretations rightly takes very seriously the distinction between reflected and reflecting perceptions. Bricke puts the distinction in terms of a distinction between the perceptions belonging to self as subject, and those belonging to self as observer; Ainslie puts it in terms of primary versus secondary ideas (see T 1.1.1.11; SBN 6). Penelhum speaks simply of our past perceptions and our present ideas of them. Bricke thinks that if the succession of reflected perceptions are all past when the succession of reflecting perceptions occurs, as Hume says at T App. 20; SBN 635, then "the self-observer cannot link the subjects's past and present perceptions." In such a case he asks, "How does my *present* self come into the picture?" Ainslie develops this suggestion and restricts the scope of the problem to cases of philosophical reflection on the mind. He argues that Hume gives no explanation how we come to believe that the secondary ideas are united with the mind. It is only the primary perceptions we are reflecting on that we represent as being united. Penelhum supports Ainslie by contending that Hume cannot explain "the synchronic identity of the mind that ascribes diachronic identity to its own past".[36] This interpretation assumes that the mind successfully represents the primary perceptions as united. If my interpretation is right, then the assumption is false. However, for the purposes of argument, suppose the assumption is true. Then Hume can indeed give the required explanation. When we represent the primary perceptions as united, we are not thinking about the secondary ones. Once we think about them (by means of tertiary perceptions), we can represent them as united with whatever other perceptions we are thinking about.[37] If any time we think about higher-level perceptions, we perceive them as united with the mind, then we can by induction conclude that every perception is united with the mind, even those we aren't thinking about. Hume gives a similar account of how we come to believe of all qualities that they are dependent on substance (T 1.4.3.7; SBN 222). If the problem raised by the fourth group were the one that Hume considered in the Appendix, it seems unlikely he would have been stymied.

Pitson argues compellingly, contra Ainslie, that the problem Hume raises in the Appendix is with his account of the vulgar belief in personal identity, not just a philosophical belief. The parallels are so striking between the account of acquiring the idea of a simple, identical self and the 1.4.2 account of the vulgar's acquiring belief in body, that the former must be an account of acquiring a vulgar belief as well.[38] It is to that account that

Hume explicitly refers in the Appendix when raising the problem. I must qualify slightly my agreement with Pitson when he argues that the vulgar have only *de re* beliefs about perceptions in the acquisition process. I do agree that, when coming to the idea of self, the vulgar do not attribute identity to successive perceptions, thought of as perceptions in accordance with the philosopher's doctrine of "double existence" (T 1.4.2.46; SBN 211). Nonetheless, I contend that the vulgar do think of them as perceptions to the extent that they pay attention to "the action of the mind in the meditation" (T 1.3.8.16; SBN 106)—something they ignore when coming to believe in body.

Pitson's own careful interpretation of the Appendix problem is as follows: Because of discontinuities, especially periods of unconsciousness, in our train of past perceptions, the perceptions exhibit insufficient coherence to cause us to attribute to them a continued, identical existence.[39] The telling objection is made by Pitson himself. We are not aware of complete gaps in consciousness. Further, our thoughts often tend to take up where they left off before the gap. So gaps in consciousness might be appealed to as a philosophical challenge to our idea of personal identity once the idea is acquired, but they cannot be a problem for Hume's account of the initial acquisition of the idea by the vulgar. Further, Pitson's interpretation doesn't make "contradiction" the right word for Hume's problem, as Pitson also seems to acknowledge.[40]

Conclusion

There are three important criteria, generally agreed on by commentators, for judging the success of an interpretation of Hume's Appendix discussion.[41] First, can it make sense of Hume's claim that his prior account is inconsistent? Second, does it explain why he says what he says and why he does not say what he does not say, and in doing so show that the problem is one Hume himself is concerned with? Third, does it explain why the problem comes up only in the course of considering the self and the idea of the self? My interpretation answers these questions as follows:

First criterion: Hume begins the Appendix worried that he has not escaped "contradictions, and absurdities" such as those that plague theories of the material world. He ends hoping that someone will eventually be able to reconcile "these contradictions" (T App 10, 21; SBN 633, 636). Notoriously Hume says, "In short there are two principles, which I cannot render consistent; nor is it in my power to renounce either of them, viz. *that all our distinct perceptions are distinct existences*, and *that the mind never perceives any real connexion among distinct existences*" (T App. 21; SBN 636).

The standard move is to say that Hume does not mean that these principles are inconsistent with each other. Rather he is saying that they are jointly inconsistent with the rest of his "defective" account. I have made this move as well, have shown the contradiction the account entails, and have explained why it is these two principles between which Hume feels he must choose to resolve it.

Second criterion: My interpretation stays very close to the text of the Appendix passage. I show how everything he says there directly contributes to the inconsistency. I do add, as every interpreter must, something not explicitly mentioned by Hume in order to arrive at an inconsistency. It is a virtue of my interpretation that I add something Hume takes to be obvious and leaves implicit in another important passage, viz., the accurate representation of our perceptions by consciousness. Thus, though Fogelin is right that the question "Why didn't he say so?" embarrasses the proponent of any Hume interpretation here, I think I can be the least embarrassed. Properly understood and given an assumption he took to be obvious, Hume did say so.

Third criterion: Why does the problem come up only when reviewing his account of our belief in personal identity?[42] It is because only there does the accuracy of consciousness come into play. Only when we are considering that of which we are most intimately conscious—our perceptions as such—is it implausible that we make mistakes. Only there, is the fiction of taking many related things in succession to be identical, implausible.

It may well appear, despite what I have said, that the accuracy of consciousness would hamper other instances in which we "ascribe an identity to related objects"—notably in our acquisition of the idea of continued and distinct existence. When faced with interrupted yet constant perceptions, "we are not apt to regard these interrupted perceptions as different, (which they really are) but on the contrary consider them as individually the same, upon account of their resemblance" (T 1.4.2.24; SBN 199). If, however, "the perceptions of the mind are perfectly known" by consciousness (T 2.2.6.2; SBN 366), and if (as seems plausible at first glance) all ideas of perceptions are consciousness's ideas, then the perceptions are too well known for us to fool ourselves about them. Only if distinct perceptions were somehow also identical could consciousness's ideas represent distinct perceptions as identical. Thus, the problem that I have said is confined to the idea of an identical self, seems more widespread.[43]

However, though all perceptions are perfectly known to us by consciousness, not all ideas of perceptions are consciousness's ideas. Consciousness's ideas are of perceptions *as perceptions in the mind*, that is, they include attention to the "action of the mind in the meditation, that certain *je-ne-sçai-quoi*," that goes with each perception (T 1.3.8.16; SBN 106). Other ideas—presumably ideas of the imagination—can fail to be considerations of perceptions as such. They can be used to regard perceptions as, for instance, "a hat, or shoe, or stone" (T 1.4.2.31; SBN 202). Consciousness's ideas play a role in the genesis of the imagination's idea of continued and distinct existence. They signal the obvious distinctness of the perceptions that are being regarded as identical. However, the imagination turns to subterfuge rather than admitting defeat. It will "disguise" the interruption "by supposing a real existence, of which we are insensible" (T 1.4.2.24; SBN

199). In other words, the imagination proceeds "upon the supposition that our perceptions are our only objects, and continue to exist even when they are not perceiv'd" (T 1.4.2.48; SBN 213). To accomplish this, the ideas of the imagination must presumably fail to attend to the "action of the mind" when representing a perception separated from the mind, in order to avoid thinking of the perception as "a real perception in the mind" (T 1.3.8.15; SBN 106). Thus, "consciousness is nothing but a reflected thought or perception" where the reflection includes attention to the action of mind (T App.20; SBN 635), whereas other ideas of perceptions are reflected thought or perception but with inattention to the action of the mind.

So, when acquiring the idea of an identical self, we regard our perceptions as perceptions. No other ideas besides consciousness's ideas are appropriate. And consciousness's ideas do not misrepresent. It is this difference between the cases which Hume overlooked in the body of the *Treatise* and becomes keenly aware of in the Appendix.

Both in the case of the idea of external existence and in the case of the idea of the self, an identity is ascribed to distinct perceptions. In the former case, the inattention of the imagination to the "action of the mind" enables it to feign a continued existence of something we are not intimately conscious of. However, in the case of the self, we are supposed to be intimately conscious of everything about it. With consciousness's accurate ideas, there can be no inattentiveness or feigning. If identity is to be ascribed by consciousness's ideas, it must, paradoxically, really hold between the distinct perceptions themselves.

Hume's only recourse is to "plead the privilege of a sceptic, and confess, that this difficulty is too hard for my understanding." Nonetheless he holds out hope that the difficulty is not "absolutely insuperable" (T App.21; SBN 636). Why did our skeptic not rest content with this further skeptical result? Perhaps it is because he had high hopes for his Pyrrhonian Empiricist "science of man." By "the application of experimental philosophy to moral subjects" (T Intro.7; SBN xvi), by "careful and exact experiments" to determine the "powers and qualities" of the mind (T Intro.8; SBN xvii), he hoped to achieve even greater glory for himself and his native country than Newton had by this approach to "natural" subjects. Penelhum suggests that the accuracy of consciousness is a precondition to Hume's science of man.[44] Perhaps Hume hoped that the accuracy of consciousness and the lack of contradiction in its domain would enable him to produce a science that would convince the world—a permanent contribution to learning.[45] Although contradiction cannot unsettle natural or firmly established beliefs for long, it would be an impediment to a stable and lasting acceptance of novel results of "refin'd reasoning." Hence, perhaps, the hope that someone "may discover some hypothesis, that will reconcile these contradictions" (T App.21; SBN 636). As I have shown, that hypothesis would have to be a solution to Hume's difficulty concerning identity.[46]

6 Systematic exposition of Hume's difficulty

Hume's difficulty concerning identity is not just a difficulty for Hume. His own response to it depends on his assumption that identity is identity through time and on his idiosyncratic views about duration and steadfast objects. The difficulty itself, however, is independent of these Humean assumptions. It depends only on the fact that we can be unsure whether or not two things are identical, and can consistently imagine them—the same them—either way. One obstacle to appreciating Hume's difficulty is that it can be mistaken for an unfortunately phrased version of Frege's famous puzzle. But whereas Frege's puzzle is, "How can an identity sentence be both true and informative," Hume's difficulty is, "How can we represent there as being some things that are perhaps numerically identical and perhaps numerically distinct?"[1]

Distinguishing these problems requires distinguishing what I have called intentional object, *what a representation represents there as being*, from what I have called intended object, *what there is which the representation represents.* Intentional objects depend for their existence on being represented, whereas intended objects (with only special exceptions) do not.[2] It is a distinction that theories of representation cannot do without, though it is often left implicit. Whether our representations are like pictures or sentences or something else, still we generally represent there as being things with certain characteristics. The *Mona Lisa* represents there as being a woman with a vague smile. The Sherlock Holmes stories represent there as being an especially observant detective. It may be that nothing is what there is which a representation represents, as in the case of a fictitious representation. Or it may be that what there is which a representation represents differs in important ways from what it represents there as being, as in the case of an inaccurate news report or an optical illusion. For example, an innocent person can be represented as being guilty; a straight stick in water can be represented as being a bent stick.

Note that there can be this differing even for true or correct representations. For example, suppose we "think with the learned and speak with the vulgar" by saying "The sun rises and sets each day."[3] Insofar as we are thinking with the learned, we recognize that the situation that makes the sentence true is one in which each day the earth rotates while remaining approximately the same distance from the sun. We may even, in special learned contexts, use the sentence to convey just this truth condition. However, insofar as we are speaking with the vulgar, we are representing there as being a situation in which each day the sun is close to the earth, then further away, then close again. That is how it looks and that is what our ordinary way of speaking conveys, even if that is not what makes the sentence true. Perception of color is another example. When I visually represent that my cup is blue, and I do so as correctly as the case admits, what there is which I represent is, say, a complex relation between me, a region on my cup, and viewing conditions.[4] But what I represent there as being, is my cup with a simple property, not in a relation at all.

It may well be, then, that representations which we successfully use, and which thereby deserve to be called true or correct, portray a world somewhat different from the world as it really is. The representations may portray features somewhat different than the real world features they track, though the tracking makes them true or correct. If so, then an account of what there is which we represent is not necessarily an account of what we represent there as being. Thus a difficulty concerning the one may not be resolvable by appeal to the other.

Frege argues, for instance, that sometimes the semantic value of "Twain" is a person—as in "Twain is Clemens"—and sometimes the semantic value is a way of picking out that person—as in "It is informative that Twain is Clemens." But either way, the semantic values are, according to Frege, objective entities that exist independent of being represented. So they are intended objects.

On the other hand, in Hume's compressed discussion of the idea of identity, the main concern is with what the idea represents there as being. The question is, what does an idea expressing that something is numerically identical with something, represent there as being? Now Hume says that a proposition about identity cannot be known intuitively. That is, the ideas that compose it do not, of themselves, determine or make it obvious whether it is true or false.[5] So entertaining such a proposition is compatible with realizing that, for all we know, its intended objects may be one and the same thing and, for all we know, may be two distinct things. So the question becomes, what do we represent there as being when we realize that for all we know some things are identical and for all we know they are distinct?[6] How can we have intentional objects neutral between being identical and being distinct?

Even if, *pace* Hume, we do not always represent there as being such a situation when we represent identity, we at least sometimes do. When you are uncertain whether your amiable neighbor is that dour German professor, you can imagine them in a way that leaves the question open. This way of imagining them allows you to go on to imagine them as identical, yet also allows you to go on to imagine them as distinct. How do you do this?

It is not that you imagine your amiable neighbor as perhaps sometimes dour, Germanic, and professorial, yet also as perhaps not ever so. What you are unsure about is not simply your neighbor's characteristics, but is rather the identity of neighbor and professor. Your neighbor could sometimes have all those characteristics and yet be distinct from the professor you have in mind. So how do you imagine there to be persons perhaps identical and perhaps distinct?

Hume concludes that there is no way consistently to so imagine. However, his account of inconsistently imagining it turns out to require an account of consistently imagining it. Can a consistent account be given? It seems difficult to do so, as I will explain.

Because Hume's difficulty is unfamiliar and hard to bring into focus, I will give three characterizations of it. Then I will show that Hume's difficulty cannot be given a Fregean solution and so differs from Frege's puzzle.

The approach in this chapter will be very different from that in its predecessors. I will leave aside Hume interpretation and merely try to give an exposition of the difficulty that underlies his attempts to render the idea of identity as that of "a medium betwixt unity and number." Philosophical problems are real, not in the sense that they exist independently of thinkers, but in the sense that there is always more to them than is captured by the formulation of any, or even all, thinkers. Accurate and penetrating as it is, Hume's extremely compressed formulation of the difficulty concerning identity, coupled with his idiosyncratic assumptions about steadfast objects and duration, serve more to obscure than to illuminate the problem. Further and more systematic exposition is called for, especially given the subtlety of the problem, the entrenched ways of neglecting it, and the unpracticed clumsiness of our language when forced to try to characterize it.

First characterization

What do we represent there as being when we realize that something and something are for all we know numerically identical and for all we know numerically distinct?

Do we represent there as being something which is perhaps identical with itself and perhaps distinct from itself? Surely not. That would be to represent there as being a situation in which only the identity is epistemically possible. In that situation, the epistemically possible distinctness is precluded by the absurdity of something being perhaps distinct from itself. The absurdity of what we are representing there as being would be too clear.[7]

Well then, do we represent there as being two distinct things which are perhaps distinct and perhaps identical? Surely not. That would be to represent there as being a situation in which only the distinctness is epistemically possible. In that situation the epistemically possible identity is precluded by the clear absurdity of distinct things being perhaps identical. If that absurdity is not as obvious as the preceding one, then consider it this way:

If we represent there as being actually distinct things which are perhaps identical, then we represent there as *perhaps* being something which is actually distinct from itself.

So we cannot represent there as being one thing; nor can we represent there as being two. But what alternative is there to representing one or representing more than one?

Do we alternately consider the alternatives, in turn representing there being one and then more than one? No. This would result in the problem just mentioned for each possibility, respectively.[8] Do we consider both alternatives at once? No. That would just compound the problems.

Do we represent there as being something and something and neither represent them as identical nor as distinct? This is no help. It doesn't matter whether we represent them as identical, or distinct, or neither. The problem comes with representing *them* in certain epistemic possibilities. That is, we have to represent there as being something and something that are perhaps identical (and so one single thing) and perhaps distinct (and so two distinct things). That means that we have to represent there as *perhaps* being one single thing which is the same thing(s) as what are perhaps two distinct things. But this is to represent there as perhaps being something which is perhaps distinct from itself.

Let me restate this with an example. Suppose we represent there as being a man, Cicero, and a man, Tully, such that perhaps Cicero is Tully and perhaps he is not. Call the epistemic possibility, which we represent there as being, of their identity, "possibility 1." Call the epistemic possibility, which we represent there as being, of their distinctness, "possibility 2." We represent there as being epistemic possibility 1 in which there exists something which is the same Cicero and the same Tully as the distinct Cicero and Tully in epistemic possibility 2. So we represent there as perhaps being something (i.e., Cicero/Tully in possibility 1) which is the same thing(s) (i.e., the same Cicero and the same Tully) as what are perhaps distinct things (i.e., Cicero and Tully in possibility 2).[9] But to represent there as perhaps being something identical with what are perhaps distinct, is to represent there as perhaps being something which is perhaps distinct from itself.

The absurdity of this situation can be rendered more striking with the help of an assumption not used above. Assume that for these cases, we can represent as *being* whatever we can represent as *perhaps* being. If we can represent there as perhaps being something that is perhaps distinct from itself, then we can represent there as being something that is perhaps distinct from itself; so we can represent there as being something distinct from itself. But this is to represent there as being a situation involving a clear absurdity.[10]

So in none of the alternatives do we represent there as being a situation free of clear absurdity, in which there is perhaps identity and perhaps distinctness. So what *do* we represent there as being when representing there as perhaps being and perhaps not being identity? Not one and the same thing, not two distinct things, not things that are neither one and the same nor two and distinct, and not things that are both. But there is no other alternative. So how can we represent there as being things perhaps identical and perhaps not? This is Hume's difficulty concerning identity.

Second characterization

Perhaps another approach will clarify the difficulty. Consider first a putative worry concerning what we represent there as being when we are unsure of whether or not something has a given property. It is a puzzle easily solved, perhaps, but which is preparation for a more difficult one. Suppose someone is not sure whether or not a given cup is blue. He says "For all I know the cup is blue and for all I know it isn't." Regardless of what makes this sentence true, he has represented there as being the same cup in different epistemically possible situations, such that it is blue in one and not in the other. The putative worry is, Why hasn't he represented there as being a contradiction? How can the blue cup which he represents there as being in one situation, be represented as being both the same cup and not blue in the other? Apparently he has represented there as being a cup that is both blue and not.

The problem is not that people cannot represent there as being contradictions. Of course they can. The problem is that he *has not* represented there as being a contradiction, and we need to explain why not.

Any answer is going to involve the fact that our way of representing there as being identity between situations is compatible with the thing having a property in one situation and lacking it in the other. Our way of representing it only precludes having and lacking a property in the same situation, assuming we are representing consistently. Just as we can represent there as being things altering through time, so we can represent there as being things "altering" across epistemic situations.

Hume's difficulty is much like the previous puzzle, except that it concerns a case in which someone is unsure whether or not something is identical with something else. Suppose someone is unsure whether Mark Twain is Samuel Clemens. She says, "For all I know Twain and Clemens are identical and for all I know they are distinct." Regardless of what makes this sentence true, she has represented there as being the same Twain and the same Clemens in different epistemically possible situations such that they are identical in one and not in the other. The worry is, why hasn't she represented there as being a contradiction?

The problem is not simply that she has represented there as being the same Twain and Clemens that have a relation in one situation which they lack in another. Identity across such situations is not in itself any more problematic than it was in the blue-cup case. The problem concerns identity across situations specifically where identity is had in one and lacked in the other.

To represent Twain and Clemens as identical in one epistemically possible situation is to represent there as being a single person in that situation who is both Twain and Clemens. To represent them as distinct in another situation is to represent there as being two persons in that situation one of whom is Twain and the other of whom is Clemens. Thus, to represent the Twain and Clemens in each situation as the same Twain and the same Clemens, is to represent there as being someone in one situation who is the same Twain as one person in another situation as well as the same Clemens as a distinct person in that other situation. But how does the representer represent there as being one person who is the same person or persons as each of two distinct people? She must represent there as being one person who is two people, that is one person who is not one person.

Before, with the cup example, the contradiction could be resolved simply by letting the difference in situation insulate the having of the property from the lacking of it. However, in the present case, the identity of the things between the situations cuts through the insulation. Representing there as being distinct Twain and Clemens in one situation and identical Twain and Clemens in the other is inconsistent with representing there as being the same Twain and the same Clemens in both.

Third characterization

How does the difficulty come about? What features of representing-as are responsible for it? I will try to answer these questions, and in doing so give a third characterization of the difficulty. However, the difficulty is, I think, independent of the particular characterization I will give of representing-as.

A helpful way to think of representing there as being, is to think of it as imagining—though without the connotations of using images or of creating fiction. For some this may gut their concept of imagining, so I will also appeal to an analogy: Generally, what one represents there as being—things existing and having various properties and relations—can be described with a sequence of sentences.[11] So to a great extent, representing-as is like telling a story. Let the "story" be the situation the sentences portray, and the "characters" be the things in the story.[12] For example, representing there as being Socrates with a churlish manner is like telling a story about a character, Socrates, in which he has a churlish manner. We can do this whether or not there really is someone the character is based on or is like. The sentences used to tell a story do not report facts; they generate the story. Suppose, however, we know that Socrates exists and has a churlish manner. Nonetheless, even true sentences, insofar as they are used to tell a story, are not used to report facts. Rather they are used to characterize characters.

Characters are the same in a story just in case they are represented as being the same. Whether or not the real people the characters are based on are the same is not relevant. Presumably, in a story, anything true of a character is true of any numerically identical character.

Suppose we know the above things about Socrates but we do not know whether or not he is wise. For all we know he could be either. Suppose we consider the relevant alternatives. How would we represent there as being ways Socrates might be beyond what we actually know? It seems to me that we need to imagine them. And that is all we need to do. That is, in terms of the analogy, we tell alternate stories about the same characters keeping constant what we know. So for any story for which we consider the relevant alternatives, "Perhaps p" is true in it if and only if "p" is true in some alternate story.[13] So for example, in some such story Socrates exists, has a churlish manner, and is perhaps wise, if and only if, in an alternate story Socrates exists, has a churlish manner, and is wise.

What is it to keep constant what we know? It is to use sentences in the telling of the story that are known to be true, and to use these very sentences, or equivalent ones, in the telling of the alternate stories. Of course, as I have said, when used to tell a story these sentences are not used, for some real situation, to represent that situation. Rather they are used to represent there as being a situation. Insofar as they are used in this way, they are not about, say, the real Cicero and Tully, but rather generate the characters Cicero and Tully. Only characters generated by sentences known to be true need to be shared in alternate stories.

It is important not to think that every sentence in a story is somehow known in or according to the story and so is known. Nor can more things be known than in the original story. The point of alternate stories as I have described them, is to flesh out alternate ways things might actually be beyond what is really known. It cannot be that imagining how things actually are beyond what we know, includes our knowing them.[14] Thus, what is perhaps true in a story is relative to the perspective of the storyteller who knows some things about the real world and is not relative to the perspective of what is known according to the story (if anything).

It follows from the above characterization that any two stories with the same characters and that keep constant what we know, are alternates to each other. Thus, the relation of being an alternate story to, is reflexive, symmetric, and transitive.[15]

This simple account of representing there as perhaps being, yields four principles that can be used to generate Hume's difficulty.

P_1. Locutions like "We represent there as being" can be replaced by locutions like "In a story", and vice versa.

P_2. For any two stories, if they share the same characters and keep constant what we know, they are alternates.

P_3. For any story for which we consider the relevant alternatives, something is perhaps true in that story if and only if it is true in some alternate story.

P_4. In a story, anything true of a character is true of any numerically identical character.

The difficulty arises as follows:

1.	We know that Cicero exists and Tully exists.	Given.
2.	We represent there as being Cicero and Tully who are perhaps numerically identical and perhaps numerically distinct, and consider the relevant alternatives.	Given.
3.	In a story, A, Cicero and Tully exist and are perhaps identical and perhaps distinct.	By (P_1).
4.	In an alternate story to A, B, Cicero and Tully exist and are identical.	By (P_3).
5.	In an alternate story to A, C, Cicero and Tully exist and are distinct.	By (P_3).
6.	Story B and Story C are alternates.	By (P_2) and ($_1$).
7.	In Story B, Cicero is perhaps distinct from Tully.[16]	By (5), (6), and (P_3).
8.	In Story B, Tully is perhaps distinct from Tully.	By (P_4).
9.	In an alternate story to B, Tully is distinct from Tully.	By (P_3).

The reader may be queasy about the move from (7) to (8), given notorious problems of substituting into opaque contexts. But these contexts are not opaque. The storyteller is in charge of which characters exist and are identical within and between stories. If there is ever a situation in which names function as logically proper names—unequivocal and their use being sufficient for there being a named—it is in normal storytelling. The use of the name in telling a story helps generate the character, and the fact that all further uses name the same character is up to the storyteller.[17]

This is enough to get Hume's difficulty. I first proceed, however, to derive another way I put things previously.

10.	In Story B, something is perhaps distinct from itself.	Existential Generalization.
11.	In Story A, there perhaps exists something perhaps distinct from itself.	By (P_3).
12.	We represent there as perhaps being something perhaps distinct from itself.	By (P_1).

The foregoing was the contentious part of Hume's difficulty. The rest of the difficulty proceeds as follows.

13.	Premise (1) does not entail that we represent there as being something distinct from itself, nor that we represent there as perhaps being something perhaps distinct from itself.	Given.

And so (13) is both true and false.

I have said that it is up to the storyteller which characters the story being told is about. In general, one is to a great degree in charge of what one represents there as being. So if one represents there as being Socrates, who is perhaps wise, then one does not have to worry whether Socrates in the one alternate story is really the same Socrates as Socrates in the other. Nor does one have to worry whether being wise is really the way one is representing Socrates as being in the alternate story. One is in charge of these things. There can be lapses, of course, as in all things human, but these would be flukes. It is as out of place to wonder whether we are right about these things as to wonder whether a storyteller making up a story is continuing to talk about the same character if he says he is, or whether he is mistaken in the characteristics he attributes to the character.

Not that everything about a story is in our control. We cannot merely by fiat make false stories true. We cannot by fiat make an inconsistent story consistent, though we can fail to recognize the inconsistency.[18] This happens in Hume's difficulty. We are surprised that we seemingly cannot represent there as being a situation properly described as in (2) without it also being properly described as in (12) and so involving absurdity.

Hume thinks there is an inconsistency hidden in (2). If we disagree with Hume, the difficulty is how to remove the one that has apparently been revealed.

No Fregean solution

Frege's successful solution to his own puzzle is not a solution to this one. A Fregean solution would make two assumptions. *First assumption*: Even in our imaginings, we can only represent there as being a given particular by representing there as being something picked out by a given sense, that is, by means of a uniquely applicable generalization. For example, to represent there as being Socrates, we represent there as being something Socratic. Let being Socratic be having the characteristics or relations by appeal to which the sense of "Socrates" picks out some one thing. *Second assumption*: The closest we can come to representing there as being numerical identity between possible situations is to represent there as being something in one and something in the other that each satisfy the same uniquely applicable generalization. For example, to represent there as being Socrates who is perhaps wise, we tell a story in which there is something Socratic, and tell an alternate story in which there is something Socratic and wise. Given these assumptions, the Fregean could apparently solve Hume's difficulty as follows: (2) entails merely that there is one alternate story in which something is both Ciceronian and Tullyesque, and another alternate story in which something is Ciceronian and some other thing is Tullyesque. The only way in which Cicero in one is "identical" to Cicero in the other is that they resemble with respect to being Ciceronian. Likewise for the Tullyesque. Thus, in no story is something perhaps distinct from itself. At most something perhaps resembles distinct things in different respects.

For the Fregean solution to fail, it is enough that either of the assumptions fail. However, both do. The sources of trouble arise from two facts about representing-as:

(i) We can represent there as being a particular without representing there as being something that satisfies a uniquely applicable generalization.
(ii) We can represent there as being something which does satisfy a uniquely applicable generalization but perhaps does not.

The first of these facts may be illustrated by the following story: There once was someone who lived somewhere. There were lots of others who lived lots of other places, too. Everyone wanted him, and also everyone else, to do a specific thing. Like the rest, perhaps he did it and perhaps he didn't. We'll never know. This story can make us anxious for a sequel without containing any uniquely applicable generalizations.

An example of the second fact is the following: Suppose we know only that there is someone Socratic but not who it is. We can tell a story according to which a certain someone is Socratic though perhaps is not. The fact that we can do so depends on our ability to represent there as being the numerically identical someone independently of how he is picked out.

The Fregean may protest that we keep characters constant through imagined vagaries of one sense by holding fast to another. For example, we can imagine someone Ciceronian to be perhaps non-Ciceronian only by imagining someone Ciceronian and Tullyesque such that perhaps there is someone non-Ciceronian yet still Tullyesque. However, this isn't right. We can imagine that very someone we started out with to be both non-Ciceronian and non-Tullyesque, as well.

Thus the Fregean solution fails.[19] Mediation by uniquely applicable generalizations is superfluous. The storyteller is directly in charge of what characters exist in a story, and which characters are numerically the same in alternate stories.[20]

To the best of my knowledge, Hume's difficulty concerning identity remains unresolved.[21]

Conclusion

I have presented a sustained argument that Hume's idea of identity is designed to circumvent a central difficulty with identity by appeal to his special views of time—a difficulty independent of those special views and which remains unresolved. The argument is long and involved, so a brief recounting may help to bring it into view as a whole. After that I will conclude by revisiting the Principle of Defensibility and then considering Hume on identity through time in the context of his skeptical approach to inquiry.

Overview

Chapter 1, "Interpreting Hume as Metaphysician and Skeptic," argues first that Hume's famous exhortation to commit various books to the flames has been too loosely interpreted. He does not mean to condemn every work that contains any metaphysics. Instead, Hume is dramatically, but merely, condemning books that both purport to be contributions to knowledge and yet are inconsistent with the results of proper mathematical or experimental reasoning. The bulk of the chapter defends an interpretation of Hume as a Pyrrhonian Empiricist. In other words he is a radical skeptic who is willing to go along with views about the world that seem forced upon him—especially by the senses—without endorsing them as ultimately true and without thinking he has any ultimately good reason to believe them. On his view, belief is a mere enlivening of ideas by causal mechanisms with no apparent connection to truth. One mechanism for enlivening beliefs is tradition, so even such a skeptic could do metaphysics that involved reasoning from traditional metaphysical assumptions. Another mechanism is taste. Where traditions conflict or fail, such a skeptic can acquiesce in metaphysical views that, after some thought, strike him to be more likely to become established. In this way, the skeptical metaphysician is like an art critic trying to decide which new works are likely to become classics, as Hume describes in "Of the Standard of Taste."

Chapter 2, "Moments and Durations," begins discussion of Hume's account of time, which is essential to his account of identity. Understanding Hume on identity requires understanding him on duration, moments, and steadfast objects. In the first section I consider how time—what he also calls duration—is an abstraction from the successions we experience, such as a succession of notes played on a flute. In other words, the idea of time is a general idea of any succession of objects just insofar as it is a succession. Time is not for Hume a dimension or container of events. Some consequences of this view are that time has parts, and that all and only successions have duration. A further consequence of this view is that Hume should think that the idea of a moment of time is the general idea of any member of a succession just insofar as it is a member of a succession. Temporal location in general is place in a succession in general. The particular temporal location of something is its place in a particular succession.

In the next section of Chapter 2, I show that for Hume moments are not composed of briefer moments. They are partless parts of time. Consequently, moments are single things. The contrasting view is that every moment of time has moments as parts, and so time is infinitely divisible. Hume's arguments against this contrasting view have been regarded with some contempt by important commentators such as Flew, Fogelin, and Laird. I show that the arguments against Hume are confused and show a lack of understanding of the conception of a line proposed by the mathematician Georg Cantor. Then I give Hume's two arguments that moments of time have no parts. The first argument, borrowed from a writer named Malezieu, is that anything divisible is really many things, not one single thing, but only single things exist. Thus, the only moments that exist are indivisible. Hume's second argument is that if the present moments were divisible into successive parts, then a contradiction would be true: All those parts would be present and yet only one of them would be present.

From these arguments, it also follows that there are temporal simples, which are single things, and that something with duration is really many temporal simples in succession.

Chapter 3, "Steadfast Objects," explains that for Hume there are objects without duration that coexist with successions of objects. Lacking duration and, therefore, parts, these steadfast objects are single things, as are any partless things, even though the successions they coexist with are themselves multiplicities. So some temporal simples coexist with successions. Failure to appreciated the strict unitariness of steadfast objects has hamstrung previous attempts to understand Hume's account of identity, for instance by Stroud, Green, Bennett, Price, and Waxman. A consequence of Hume's view of moments is that a steadfast object exists at a single lengthy moment. A single moment can coexist with a succession of moments without being composed of them. Because this view feels contradictory, I formalize it in order to show that it is consistent. I then respond to an objection based in Hume's claim that succession is the essence of time, which seems to rule out coexistent moments. Given Hume's definition of time in general as succession in general, his claim certainly makes sense. However, when considering the particular times of particular objects, his account requires that the particular moments of coexistent objects coexist.

Another reason we have trouble accepting coexistent particular moments is that the mind naturally comes falsely to think that steadfast objects have duration, according to Hume. We confuse coexisting with all members of a coexisting succession (as a steadfast object does) with successively coexisting with each member of a coexisting succession (as something with duration might). This confusion is crucial to preparing us to acquire the idea of identity. It also enables us to imagine that there is a perfect standard for duration that applies to everything in time. Thus it becomes unnatural for us to believe that some things occupy moments that coexist with the successive moments of other things.

Chapter 4, "Identity," shows how, for Hume, the idea of identity is the fictional idea of a steadfast object with duration. This currently strange way of thinking of identity is the result of two considerations. First, Hume is assuming that paradigmatically questions about identity concern identity through change, even if only change in time. Second, Hume is trying to be true to a difficulty that he sees at the heart of the concept of identity. The difficulty is this: We can be unsure about an identity, that is, we can think of two things while leaving it open whether or not they are identical. Further, we can conceive that both alternatives are possible, that is, we are able to represent them as being perhaps identical and yet are able to represent the very same things as being perhaps distinct. This concern with ignorance about an identity puts people, notably Bennett, in mind of Frege's famous puzzle how a true identity sentence can be informative. However, I show that Hume has something else in mind. Seeing the difference requires distinguishing (i) what there is which an idea represents and (ii) what an idea represents there as being. The former is like what Ayers calls "the real object" and the latter what he calls "the intentional object of thought." I show why his terminology, as Ayers explains it, is less helpful here than my initially strange formulations. While Frege is concerned with what there are to serve as senses and referent of our terms, Hume is concerned with how we represent there as being things that are perhaps identical and perhaps distinct. In other words, he is concerned with how we represent there as being "a medium betwixt unity and number." Further evidence that Hume was not clumsily presenting a Fregean puzzle is that Fregean concerns had already been fairly clearly formulated by Berkeley and Locke. Hume could easily have supplied a semantic solution based on their work if it were a semantic puzzle he was concerned with.

The best Hume can think of to explain how we represent there as being a "medium betwixt unity and number" is to explain how we represent a steadfast object recognized as such, as having duration recognized as such, while trying to hold together these conflicting aspects. They are conflicting in just the right way, because being steadfast entails being a single thing while having duration entails being many distinct things. What we do, he says, is alternately regard the object as one and as many, since we cannot think of it at once as both. This talk of one and many should not mislead the reader into thinking that an easy solution is to think of something as one whole with many parts. The problem at issue is how to think of the same things as perhaps one and the same with each other and perhaps many and distinct from each other. Hume's explanation at this point is remarkably dense, almost impenetrable, and only a careful familiarity with his account of time lets us find a way in. Given that the idea of identity is as he says, the characteristics we must experience in order to acquire that idea are "the invariableness and uninterruptedness of any object through a suppos'd variation in time." The first two capture steadfastness and the third captures duration.

I show that this account avoids Stroud's charge that Hume's account of identity begs the question. The charge is based on an assumption shared with Green and Bennett that perceptions are all exceedingly brief. I show that the assumption is false because for Hume there can be steadfast perceptions. I also show that the difficulty with identity Hume raises cannot be addressed by appeal to distinctions of reason. I show third that Hume's proposal tries to finesse the difficulty without resolving it, but that even the finesse won't work without a resolution.

Chapter 5, "Representing Personal Identity," concerns the obscure, much-discussed passage from the *Treatise* Appendix in which Hume finds a problem with his account of acquiring the idea of personal identity. I argue that Hume has discovered that the self has to be a real "medium betwixt unity and number" in order to represent itself as having identity. Therefore, the Appendix problem can be resolved only if the original difficulty with identity can be. The tacit assumption in the passage is that the sort of ideas used to represent the self—consciousness's ideas—are unerring. Thus, the self can only represent itself as being how it is. Since, on Hume's view, it is many perceptions bundled together, it can represent itself as being also a unitary thing only if it is also a unitary thing. The problem is that being many distinct things entails not being unitary. I collect several passages concerning the accuracy of consciousness and appeal to Penelhum's contention that assuming this accuracy is central to the method of Hume's science of man (all prosecuted within the context of his Pyrrhonian Empiricism, of course). The two possible solutions to the Appendix problem, ones Hume considers and rejects, are traditional attempts to make sense of being "a medium betwixt unity and number": (i) being distinct perceptions inhering in a simple mental substance, and (ii) being distinct perceptions with a necessary connection. Hume has rejected both as attempts to have perceptions be both distinct and identical and, so, is unable to see how to reconcile the same contradiction in his own account.

My interpretation requires distinguishing consciousness's ideas, with their requisite accuracy, from other ideas of perceptions subject to the fictions and confusions of the imagination. To do so, I appeal to Hume's distinction between ideas of perceptions considered as perceptions in the mind and ideas of perceptions considered as representations of absent objects. The former are consciousness's ideas. This focus on consciousness in the Appendix is what confines the problem raised there to his account of acquiring the idea of personal identity and keeps it from extending to, for instance, his account of acquiring the idea of the external world.

There are a huge number of scholarly, perceptive, and ingenious interpretations of the Appendix passage. I show that even the latest attempts by Bricke, Ainslie, Penelhum, and Pitson, which in very sophisticated ways build on the successes and failures of their predecessors, are not quite true to Hume. I show that my interpretation best explains the inconsistency Hume cannot solve, stays closest to the text, only brings in an assumption also left tacit elsewhere in Hume, and explains why the problem seemed to Hume only to apply to his account of the idea of personal identity.

Chapter 6, "Systematic Exposition of Hume's Difficulty," shows that the apparent contradiction Hume finds in the concept of identity is independent of his assumption that identity is identity through time and is independent of his idiosyncratic views about steadfast objects and duration. The difficulty depends only on the fact that we can be unsure whether or not two things are identical. Hume says that thinking of things as identical with each other requires being able to think of them as one yet also as many and so requires thinking inconsistently. He seems right. If we represent there as being *a* and *b* of whose identity we are unsure, then as we contemplate the alternatives we must be able to think of *a* and *b* as one and the same but also be able to think of them, the same them, as two distinct things. But this means that when we are thinking of them as one thing, we must be able to go on to think of it (that one thing) as distinct from itself. After all, if we can think of each of them as it, then we can think of it as each of them, and so as each of the two distinct things. So there seems to be a contradiction at the heart of identity. To see the contradiction, we have to focus on the fact that it is *a* and *b* that we are thinking about. It is them that we have to hold fixed when considering the alternatives, not their names nor descriptions nor evidences of them, as might be gathered from Frege or Kripke. Further, we are thinking of them alternately as identical and distinct from each other, not alternately as composing a single whole or being its many parts, as might be gathered from Lewis. The best contemporary accounts, for all their merits in solving other problems, do not address this one.

I give three characterizations of the problem and show that a Fregean solution is not possible. To the best of my knowledge, Hume's difficulty concerning identity remains unresolved. We should listen to what he says about it.

Defensibility, then skeptical inquiry

I claim to be guided by a Principle of Defensibility, seeking an interpretation that makes all of Hume's positions as defensible as possible. Yet I argue that he fails to evade his own difficulty concerning identity, which makes his account importantly indefensible in this respect. My interpretation thus seems not fully consistent with the principle. However, successful evasion by Hume would have required solution instead (Chapter 4), and I have shown that there is no recognized solution (Chapter 6).[1] Further, if Hume had a solution, he could have escaped the Appendix labyrinth concerning the idea of personal identity (Chapter 5), which he explicitly admits he cannot do. So, although an interpretation that made Hume's position even more defensible would have been preferable, it does not seem possible to provide one.

It remains to consider Hume's account of identity through time in the context of his larger project. I will argue that doing so informs an understanding of his general approach to inquiry: his Pyrrhonian Empiricism.

Hume's account of identity is part of his account of acquiring a belief in body—perceivable objects that continue to exist unperceived, such as hats and stones and shoes. As Hume explains, when an interrupted sequence of perceptions exhibits constancy, i.e., exact resemblance, invariableness, the mind naturally takes the distinct perceptions to be identical. One of the essential characteristics of objects exhibiting identity is uninterruptedness, so the mind takes the constant but interrupted perceptions to be uninterrupted. However, the interruption of the constant perceptions is too manifest to be ignored, so the mind comes to believe that the uninterrupted object continues to exist between the times it is perceived (T 1.4.2.24–44; SBN 199–210).

In a similar way, Hume's account of identity is part of his account of belief in a single, steadfast self. As Hume explains, when an uninterrupted but variable sequence of perceptions is bundled by close relation, the mind naturally takes the distinct perceptions to be identical. One of the essential characteristics of objects exhibiting identity is invariableness, so the mind takes the uninterrupted but variable perceptions to be invariable.[2] However, the variation is too manifest to be ignored, so the mind comes to believe that what is uninterrupted and invariable is a soul that connects the perceptions together (T 1.4.6.5–7; SBN 253–5).[3] Note the parallel to the case of body. In each case, one essential quality of identity first is missing, and then is supplied, by the mind.[4]

Seeing the parallel between the cases and understanding the structure of the idea of identity, helps establish that Hume is a thoroughgoing skeptic concerning our beliefs in body and in the self. One of the deepest and most enduring controversies in Hume scholarship concerns the extent of his skepticism. Influential early interpretations, such as those by Reid and Beattie, had Hume be a corrosive skeptic in the sense of being a negative dogmatist: one who denied the existence of external objects and the self. Under the influence of Kemp Smith, opinion swung the other way, seeing Hume's skepticism as limited by his naturalism. More recently, Kemp Smith's heirs, such as Wright and the New Humeans, and Garrett, have worked to refine that interpretation to account properly for the undeniable skeptical elements in Hume, as emphasized by Popkin, Fogelin, and Norton. Understanding

Hume on identity helps show that even these sophisticated refinements unduly limit his skepticism.

The idea of identity involves the false supposition that steadfast objects have duration. The resulting idea pretends to be a "medium betwixt unity and number," something that Hume says is strictly speaking impossible. One step in acquiring belief in body requires the idea of identity to be used to conceive the "gross illusion" that interrupted perceptions are uninterrupted (T 1.4.2.56; SBN 217). Likewise, one step in acquiring belief in the self requires the idea of identity to be used to conceive the "confusion and mistake" that variable perceptions are invariable (T 1.4.6.6; SBN 254).[5] That these beliefs are generated by means of "trivial qualities of the fancy, conducted by such false suppositions" (T 1.4.2.56; SBN 217) is no proof that the beliefs are false. Hume is no negative dogmatist. However, their being generated in this way gives no reason at all to believe that they are true, the naturalness of the process notwithstanding. Quite the opposite. In both cases, the process appears to be a sequence of natural errors highly unlikely to yield a true belief in the end. This appearance raises doubts about the resulting beliefs that, as near as he can tell, cannot be settled by further reflection and thus prevents Hume from regarding the beliefs as justified. In this state of doubt, Hume says, "Carelessness and in-attention alone can afford us any remedy" (T 1.4.2.57; SBN 218). It is only inattention to his grounds for doubt that allows Hume later to continue his investigations by being "diffident of his philosophical doubts, as well as of his philosophical conviction" (T 1.4.7.14; SBN 273). In continuing, he is being a "true sceptic" by yielding passive acquiescence to the way things strike him, even while aware that he cannot yet actively assent to the truth of these ways. Reading Hume as a Pyrrhonian Empiricist reconciles his skeptical and naturalistic sides without limiting his skepticism. Understanding Hume on identity helps us see that any such limitation is inappropriate.

Seeing how Hume can be a thoroughgoing skeptic with regard to granting active assent, and a constructive theorist with regard yielding passive acquiescence, can inform not only our reading of Hume, but also our own pursuit of metaphysics. It would seem folly to say that metaphysics is knowledge, given Peirce's trenchant characterization of the "*a priori* method" used in metaphysics as essentially an aesthetic matter.[6] Metaphysics is an attempt to find common ground concerning what feels true. This characterization would be a rebuke to any but a skeptic who regarded all inquiry that way. Hume is such a skeptic. He shows the way for metaphysicians who see no way to establish a connection between feeling true and being true, and who yet are moved by the beauty and profundity of their discipline.[7]

Notes

Introduction

1 Kemp Smith 1941: 474–6. Bennett 2001: Volume II, 298.
2 Flew 1976.
3 Ayers 2003: 1067.
4 Cf. Davidson 2001: xix–xx, 282.

1 Interpreting Humes as metaphysician and skeptic

1 Hume explicitly asserts only that anything infinitely divisible has an infinite number of parts, but it is hard to see why he would assert this if he did not hold the more general principle.
2 He seems also to equate "metaphysical reasonings" with abstruse reasonings at T Intro.3; SBN xiv. However, when he refers to "the metaphysical Parts of my Reasoning" in the *Treatise* in his letter of August 26, 1737 to Michael Ramsay, the list of recommended works to help in understanding them (of Malebranche, Berkeley, Bayle, and Descartes) seems to indicate a more restricted notion of metaphysics. The letter is quoted in Mossner 1980: 104, 626–7.
3 See Hempel 1959; Carnap 1967: 55.
4 See Frede's extended argument that the Pyrrhonians and earlier Academics were subsequently interpreted to be more dogmatic than they really were (Frede 1987: 201–22). Popkin thinks Hume's stated version of Pyrrhonism fits this description (Popkin 1966: 55–6). Commentators Hume relied on were, for example, Cicero, Diogenes Laertius, Montaigne, and Bayle.
5 I am relying for my remarks mainly on Sextus (2000), Frede, and Popkin. See also Norton 1982: Chapter 6. For a detailed discussion of Hume's familiarity with Sextus's works, see Fosl 1998. Note that Frede's interpretation concerning the distinction between two kinds of assent in Sextus is controversial among commentators on ancient philosophy. For a competing interpretation see Burnyeat's classic "Can the Skeptic Live His Skepticism," (Burnyeat 1983: Chapter 6). Annas hints at an account like the one I will give, but does not develop it and ends up reading Hume as a dogmatist (Annas 2000: especially 276 and 279). I am grateful for criticism and suggestions from Julia Annas and John Cooper, and discussion with John Wright and Peter Loptson.
6 Sextus 2000, Book 1, Chapter 7, Section 10.
7 Hume would have been familiar with the modern revival of such a view in Mersenne and Gassendi, its development by Anglican theologians and the Royal Society, and its expression in Boyle, Newton, and Locke. See Popkin 2003: Chapter 7, "Constructive or Mitigated Scepticism," and Van Leeuwen 1963. For

a wide-ranging history of reasoning without certainty, see Franklin 2001. A brief, clear, and helpful discussion of the varieties of ancient Academic skepticism can be found in Brittain's introduction to Cicero (Cicero 2006). I am following Brittain in distinguishing mitigated skepticism from radical skepticism, on the one hand, and fallibilism, on the other.

8 *History*: Vol. 6, Chapter 70.

9 It may be that Hume in some ways was following in the footsteps of the early Academic Carneades, if the latter is read as a "classical skeptic" in Frede's sense (Frede 1987: 201). It is along such lines that, in his introduction to Cicero's *On Academic Scepticism*, Brittain takes Clitomachus to interpret his teacher, Carneades (2006: xxvi–xxvii).

10 It is the sort of belief that might issue purely from "rationalist reason," to use part of Baier's memorable phrase, "those fantastic sceptics who shelter under the throne of rationalist reason." Note that what is fantastic about them, according to Hume, is not that they suspend active endorsement, but that they regard themselves as able to withstand the "absolute and uncontroulable necessity" by which Nature forces views on them in spite of that suspension. Hume regarded any reports of such a person who actually resisted nature, such as the account by Diogenes Laertius of Pyrrho, as false. Anyone who argued against living such a life "disputed without an antagonist" (T 1.4.1.7–8; SBN 183). Hume is likely right about this, but seems wrong to have thought there ever was a sect of skeptics that even tried not to assent to anything in any way at all (Baier 1991: 59).

11 There may, as well, be mitigation in the sense that after experiencing epistemic despair we tend passively to acquiesce with more "doubt, and caution, and modesty" in the course of enquiries that we are more likely to limit to "such subjects as are best adapted to the narrow capacity of human understanding," than we might have otherwise (EHU 12.24; SBN 161–2).

12 Popkin claims that nature pits epistemological Pyrrhonism against practical Pyrrhonism and that Hume's willingness to be led by nature even to theoretical views is the only hope for achieving quietude (Popkin 1966: 93–8).

13 Sextus 2000: Book 1, Chapter 33, Section 230.

14 Hume's restriction in this passage to belief as a result of "probable reasoning" is no real restriction since he thinks "all knowledge resolves itself into probability" (T 1.4.1.4; SBN 181).

15 The phrase "refin'd reasoning" is from T 1.4.7.7; SBN 268, and "elaborate philosophical researches" from T 1.4.7.15; SBN 273.

16 Precedence for seeking stability is found in Sextus's discussion of the preference of the New Academy for "appearances which are plausible and scrutinized and undistractable," or, as Mates translates it, "the *phantasia* that is plausible, tested, and stable" (Sextus 2000: Book 1, Chapter 33, Section 229) (Mates 1996). Loeb's book (2002) laudably explores in detail Hume's appeal to stability. We differ in that I think Hume's commitment to stability is an effect of his skepticism, whereas Loeb thinks the commitment is the cause of Hume's skepticism.

17 The case is the same with what Garrett (1997: 232–7) has usefully termed the "Title Principle": "Where reason is lively, and mixes itself with some propensity, it ought to be assented to" (T 1.4.7.11; SBN 270). Liveliness and propensity can yield, at best, ersatz epistemic warrant. There is no reason to regard them as evidence of truth. Nor is skeptical diffidence about one's skepticism a help in this regard. That our skepticism about our faculty of reason may cut against the skepticism that results from its exercise, is just more balancing of arguments preventing active assent. It is no reason to regard ourselves as epistemically entitled to some assent. The consideration that naturalness can provide no genuine epistemic warrrant opposes in general the position of Kemp Smith and his many heirs, including the New Humeans such as Wright (Kemp Smith 1941: 87; Wright 1983: 225).

18 Cf. Berkeley's distinction between, for instance, "sound as it is perceived by us, and as it is in itself," and his exclusive concern with the former. George Berkeley, "The First Dialogue," *Three Dialogues between Hylas and Philonous* (1948–57: Vol. II, 174–5, 180, 181–2).
19 I will note them when they arise.
20 Here Hume contrasts being real with being nonexistent.
21 Sextus 2000: Book 1, Chapter 33, Section 223. I am grateful to John Cooper for emphasizing this aspect of ancient skepticism.
22 I suspect that Janet Broughton's recent and subtle version of Hume's naturalism-cum-skepticism will have trouble answering Sextus' point. She appears to me to be reading Hume as experiencing what might be called doxastic akrasia. He believes things against his better judgment, and with the ironic detachment that many with acknowledged bad habits regard their vices, instead of suspending belief in any way (Broughton 2004: 548–53).
23 Wright also argues, in a different way, that the conclusion of this experiment requires assuming the existence of body, though uses his argument as part of his skeptical realist case that there is some warrant to our belief in body (Wright 1983: 52–53).
24 Thus it is improper to take Hume to be an Epicurean who regards sensation as a criterion of truth. Hume certainly can sound like an Epicurean when in the grip of the view that there are inner perceptions, as when he says, "The only existences, of which we are certain, are perceptions, which being immediately present to us by consciousness, command our strongest assent, and are the first foundation of all our conclusions" (T 1.4.2.47; SBN 212). However, such talk must be seen as passively assenting to a view he ultimately is unsure about. It is better to say that in his Pyrrhonian way, Hume emulates an Epicurean. I am grateful to John Cooper for suggesting that Hume is one.
25 The considerations against a framework of both an external world of objects and an internal world of perceptions, "*admit of no answer and produce no conviction*," as Hume famously says of Berkeley's philosophy (EHU 12.15 n. 32; SBN 155 n. 1).
26 It might seem odd to speak of the external world as it appears, forcing views upon those minds to which it appears. The appearances would seem rather to depend in some way on the views. However, for Hume, sorting out the dependencies would be beyond us. The best we can do, in giving a science of the mind, is find constant conjunctions between the world as we believe it to be and our ensuing beliefs as we believe them to be. I am grateful to Gunnar Björnsson for discussion of this point.
27 Sextus 2000, Book 1, Chapter 11, Section 23.
28 Sextus 2000, Book 1, Chapter 11, Section 24.
29 Peirce 1955: 17.

2 Moments and durations

1 As will be seen in the next chapter, a temporal complex also could have parts at coexistent moments, but that won't make sense until the next chapter.
2 I use "really" not to imply that Hume is departing from his skepticism, but to draw attention to the fact that the grammatical singularity of the sentence's subject is misleading.
3 T 1.1.7.1–10; SBN 17–22, and *Essay* 2.11.9, 2.12.1, 3.3.6–9, 3.6.32.
4 Hume, T 1.1.7.6, 1.1.7.10; SBN 19, 22; *Essay* 3.3.11.
5 Hume T 1.1.7.6; SBN 19. *Essay* 3.3.1, 3.3.6.
6 Boethius 1994: 20–5; Berkeley, Introduction to *The Principles of Human Knowledge* (1948–57: Vol. II).
7 See Winkler, Flage, and Baxter. Winkler 1989: 33, 36–8; see also his Editor's Introduction to Berkeley, *A Treatise Concerning the Principles of Human Knowledge*

(Berkeley 1982: xvii–xviii), and his "Berkeley on Abstract Ideas," (Winkler 1983: 69–73); Flage 1986: 489; Baxter 1997: 307–30.

8 These examples are drawn from T 1.1.7.2, 1.1.7.3, 1.1.7.8, and 1.1.7.14; SBN 17, 18, 21 and 23.

9 That is, I presume, word tokens of the same type.

10 Here I commend the reader's attention to Falkenstein's careful discussion of the consistency of Hume's empiricism with his appeal to manners of disposition. Falkenstein argues persuasively that ideas of manners of disposition can be arrived at by the copy principle. We just have to realize that some ideas are copied not from single impressions but rather from several disposed in a certain manner. Falkenstein 1997: 189.

11 Robert Fogelin has reminded me that there are other sorts of successions than temporal ones. However, temporal succession is here the relevant natural relation that causes the mind to collect ideas together. Explaining how exactly this works "is impossible," but we know by experience and analogy that it does (T 1.1.7.11; SBN 22).

12 Kemp Smith 1941: 274.

13 Green 1886: Section 259, 217–18.

14 I'm grateful to Chris Panza for discussion of this point.

15 Rosenberg 1993: 83.

16 Augustine 1961: Book 11, Section 15.

17 As here, all the fictions I mention in this paper are evident falsehoods: Taking something without duration to have it, taking distinct things to be identical, etc. Baier asserts that Humean fictions are not false, simply unverifiable, but this seems to be to be more an attempt to save Hume from himself, than an attempt to take him at his word (Baier 1991: 103). See also Traiger 1987: 381–99.

18 Kemp Smith 1941: 274; Falkenstein 1997: 180; Garrett 1997: 53–4.

19 Although Garrett says that space and time themselves are manners, I think our accounts of the abstract ideas of space and time are otherwise perfectly compatible. He focuses on what the ideas are, I focus here on what they are of. The ideas he describes are not simply of manners, I think, but also of simples arranged in these manners.

20 One might try to construct a Humean analogue to the view of time as a dimension or container by imagining the whole composed of all particular times, but Hume does not use "time" in this sense.

21 See also Berkeley, *Principles*, Part 1, section 130 (Berkeley 1948–57: II, 101).

22 For discussion of some of Hume's other arguments on these issues, see Baxter 1988a and Baxter (forthcoming).

23 See especially Rosenberg, Flew, Fogelin, and Laird. Flew 1976; Fogelin 1985, 1988; Laird 1931); Rosenberg 1983.

24 Bayle endorsed the infinite divisibility assumption (1991: 356, Zeno of Elea, Remark F). However, his argument presupposes the discreteness of time, which he takes himself already to have proven.

25 Aristotle, *Physics* Book 3, Chapter 6, $206^{a}8$–$206^{b}32$ and Book 8, Chapter 8, $262^{a}22$–$263^{a}3$; and *On Generation and Corruption* Book 1, Chapter 2, $316^{a}20$–$316^{a}23$ (Aristotle 1941).

26 There was widespread support for Hume's position on this issue. See Holden 2002: 3–25. For further background on the debate see Holden 2004.

27 Bayle 1991: 360, Zeno of Elea, Remark G.

28 I am grateful to Thomas Holden, Lorne Falkenstein, and Phillip Bricker for discussion of the Aristotelian objection.

29 See Flew 1976: 260.

30 Bayle 1991: 356, Zeno of Elea, Remark F.

31 Cf. Lewis 1991: 1–6.

32 See Grünbaum 1967 for a characterization of the Cantorean conception of a line.
33 Flew perhaps would have acknowledged this if he had taken the distinction between aliquot and proportional parts in the way explained by Frasca-Spada, who has found the appropriate meanings for the terms in Bayle's *Systeme Abregé de Philosophie*. Aliquot parts are ones of equal size. Proportional parts decrease in size in a certain proportion. In the example I just gave, I appealed to proportional parts; Frasca-Spada 1998: 36. Flew's misleading hints about this distinction are at Flew 1976: 264, and are developed by Newman (1995: 43–4).
34 Fogelin 1985: 27–8.
35 Cf. McRae 1980: 123; and Laird 1931: 67.
36 Fogelin 1988: 51.
37 Fogelin 1988: 64.
38 Aristotle (1941) *On Generation and Corruption*, Book 1, Chapter 2, 316ª14–317ª17.
39 For previous discussions of Hume on parts and wholes, see Baxter 1990 and 1988a.
40 Kemp Smith has noticed this commitment of Hume's, but gives no explanation (1941: 500 n. 1).
41 See also T 1.4.3.2; SBN 219 where Hume connects unity with simplicity and composition with lack of simplicity and, so, lack of unity.
42 Support comes from Bayle's claim that "All philosophers agree that the material cause is not distinct from its effect. Therefore, what would be composed of unextended parts would not be distinguished from them" (Bayle 1991: 394, Zeno the Epicurean, Remark D. In other words, something just is the parts it is composed of. See Baxter 1988b and 1988c.
43 Hume's explanation is that many related things collectively resemble a single thing (T 1.4.3.5; SBN 221). It is a feature of our psychology that we treat fairly close-knit pluralities this way. See also Baxter 2001a.
44 Cf. Bayle: "number includes essentially several unities" (1991: 381, Zeno of Elea, Remark I).
45 Flew 1976: 264.
46 See Leibniz's argument in the correspondence with Arnauld that to be real a corporeal substance must be one and indivisible (1967: 120–2). For commentary, see Baxter 1995. See also Cummins 1990.
47 Flew 1976: 265.
48 I'm grateful for criticisms by Gideon Yaffe of my early renditions of this argument.
49 Bayle 1991: 353–4, Zeno of Elea, Remark F. David Raynor directed me to Bayle's argument.
50 I am assuming that if moments are divisible, they are divisible into moments. What else?
51 See Grünbaum 1967: 45–56.
52 Bayle 1991: 354, Zeno of Elea, Remark F.
53 Another way to read Hume's argument is along the lines of Plutarch when criticizing the Stoics on time: If the present moment is divisible into successive parts, then the present overlaps with the past (the earlier parts) and the future (the later parts), so the present moment is past and future as well, or, equivalently, past and future parts of time are both present. Plutarch, *On Common Conceptions*, 1081c–1082a (2001: 285–6).

3 Steadfast objects

1 Forerunners of my interpretation that not all perceptions are exceedingly brief are McIntyre 1976: 79–88; Hirsch 1983; and van Steenburgh 1977. See also Costa 1990 and Garrett 1997: 52–5.

2 Johnson 1755: Steadfast.
3 By "steadfast," neither I nor Hume mean permanent. Something relatively short-lived is steadfast as long as it coexists with a succession of shorter-lived things.
4 He also says, "[T]he relations of *contiguity* and *distance* betwixt two objects may be chang'd merely by an alteration of their place, without any change on the objects themselves or on their ideas" (T 1.3.1.1; SBN 69). I'm indebted to Eli Hirsch for the following interesting problem: If steadfast objects can change place, and a mass of matter can be composed of shifting steadfast objects, then the mass is the same mass yet has a succession of shapes. If shape is a quality then altering shape is variation contrary to identity through time. So the same mass is not the same mass. In answer, I conjecture that for Hume shape isn't really a quality, because things with shape are not really individuals. Anything extended, rather than being "an unite," is a number of individual minimal points (T 1.2.2.3; SBN 30–1). So, except for the degenerate minimal shape, all shapes are relations holding between numerous points. They are external relations "such as may be chang'd without any change in the ideas" (T 1.3.1.1; SBN 69). So, change in shape is not contrary to the steadfastness of the shifting points. Granted, Hume calls "figure" a "quality" at T 1.4.3.5; SBN 221, but if he is to be consistent he must be speaking loosely.
5 Annette Baier drew my attention to this passage.
6 How is the passage of time felt in this case? It surely is felt, even if "scarce felt." Presumably we feel the "changes of the co-existent objects" (T 1.4.2.29; SBN 201), even though we are hardly paying attention to them. I'm grateful to James Dye for asking the question here.
7 Stroud 1977: 103.
8 Stroud 1977: 101–4; Green 1886: Sections 303, 255–6; Bennett 2001: II, 298–9.
9 Price 1940: 46–7.
10 Price 1940: 46.
11 Stroud 1977: 105.
12 Waxman 1994: 207.
13 Waxman 1994: 206–7.
14 Waxman 1994: 322 n. 11.
15 See Costa 1990: 12.
16 Bewitched by this view of time as a line, I earlier attempted to make sense of existing more than just briefly without having duration, by suggesting that a steadfast object can occupy a temporal interval without occupying its sub-intervals. See Baxter 1987: 333. This attempt, however, relied on a view that a whole is a distinct thing from its several parts. Such a view of parts and wholes cannot be Hume's, not even for temporal intervals, which, being abstractions, will have the structure of the wholes they are abstracted from. See my Chapter 2 discussion of the Malezieu argument (T 1.2.2.3; SBN 30–1).
17 I mean a vertical line anywhere along the extent of a block, not just at the ends.
18 I'm grateful to Virgil Whitmyer for some corrections.
19 Akihiro Kanamori points out in his editor's introduction that a weak simultaneity relation can be defined on Humean moments which divides them up into equivalence classes that themselves are totally ordered by the later-than relation. Presumably there would be more than one such equivalence class only if not every moment and a moment later than it both coexisted with some moment (2000: xxxvi–xxxvii).
20 For background, see Hume's discussion of the idea of a vacuum, esp. T 1.2.5.5–21; SBN 55–62. Costa gives a good summary of what Hume is up to.
21 I owe this idea of flow rate to Peter D'Alesandre.
22 See T 1.2.5.29; SBN 65. For Hume, a perception is any object of direct awareness, whether sensory awareness or some other.

23 See Williams 1951.
24 Previously, I used the terms "simple time" and "complex time" to distinguish time or duration from co-duration. However, I otherwise use "simple" to connote partlessness and here want to avoid any ambiguity. Duration (née simple time) has parts. See Baxter 2001b: 138.
25 The fact that when we *experience* a succession of resembling things we tend to regard the succession as instead "one continu'd object" (T 1.4.3.3; SBN 220), does not prevent us from forming the idea of a succession of resembling things.
26 Though I am in substantial agreement with Costa's excellent paper, I disagree with him on this point. I think at T 1.2.5.29; SBN 65 Hume is not considering a continuous hour-long experience of the object, but rather interrupted experiences. Still, Costa is right to the extent that we have to initially think of the object as steadfast, perhaps from previous experience, for the "confounding" to occur as Hume describes and result in "the idea of time without a changeable existence."
27 Cf. Lewis 1986: 202–4, 210. Consequently it is too hasty to contend, as Penelhum does, that in taking alteration to be incompatible with identity Hume is simply making a mistake (Penelhum 2000: 31).
28 I am grateful for comments, criticisms, and suggestions from Eli Hirsch, Michael Costa, Ken Winkler, Daniel Flage, John Biro, James Dye, and the audiences for earlier versions of this chapter at the Twentieth World Congress of Philosophy, Boston, Mass., 1998, the Twenty-Seventh Hume Conference, Williamsburg, Virg., 2000, and the Pacific Division Meetings of the American Philosophical Association, San Francisco, Calif., 2001.

4 Identity

1 The quoted phrase is from a sentence about causation, but Hume has just said, "'Tis the same case with *identity* and *causation*."
2 Frege 1980. As is well known, Frege's puzzle is a puzzle how substituting co-referential terms can change the truth value of sentences in general, not just identity sentences in particular. See Salmon 1986. Still it can be seen as a puzzle about identity in that it concerns the appearance that something a propositional function is true of and something that propositional function is false of are the same thing.
3 Bennett 2001: II, 298.
4 See, e.g., Lewis 1986: 202–4, 210, and the discussions spawned by it.
5 What is left of identity at the metaphysical level after Hume has denied genuine identity through time, is just what he would term "unity," in the sense of unitariness.
6 Cf. T 1.2.5.19–21; SBN 60–2.
7 With respect to (ii), see T 1.2.6.4; SBN 67 where Hume says, "Whatever we conceive, we conceive to be existent." We represent there as being what we conceive.
8 Quine 1956: 177.
9 René Descartes, Third Meditation (1984: 29). He also says, "For whatever we perceive as being in the objects of our ideas exists objectively in the ideas themselves," in *Arguments Proving the Existence of God and the Distinction Between the Soul and the Body Arranged in Geometrical Fashion*, in the *Second Set of Replies* (Descartes 1984: 114).
10 Brentano 1995: 88.
11 I'm using these terms as technical terms with little relation to the English synonym for "planned."
12 Here is an instance like the more awkward of my locutions in the cognitive science literature: "The boxes were then pushed into the child's reach, and patterns of search revealed how many objects the child had represented as being in the box" (Carey and Xu 2001: 189).

13 Pylyshyn 2001: 128.
14 Such a figure is now called a Kanizsa triangle. See Kanizsa 1987.
15 Ayers 2003: 1067.
16 What I say about this is motivated by a discussion with John Carroll and Kenneth Winkler.
17 Cf. Adams's brief characterization of intentional object (1983: 218). See also Camp 1988: 11–14.
18 Kemp Smith 1941: 474–6.
19 See also the discussion of distinctions of reason in which he treats a globe of white marble as a single thing despite its composition of indivisible parts, since again the strict multiplicity is irrelevant to the point he is making (T 1.1.7.17–18; SBN 24–5).
20 Cf. Wittgenstein (1961: 2.1–3.2) and Sellars (1979: Chapter 3).
21 This claim about ideas could be used to begin a Humean response to Fodor's objection that ideas cannot be copied from impressions because ideas have canonical decompositions and impressions do not. Hume could say that some ideas—those formed by "conjunction"—have a canonical decomposition into ideas copied from impressions. See Fodor 2003: 33–5.
22 Less clear evidence of the connection is provided by two other claims: "A single perception can never produce the idea of a double existence" (T 1.4.2.4; SBN 189) and "One single object conveys the idea of unity" (T 1.4.2.26; SBN 200).
23 For our idea of body, see T 1.4.2. "Nothing exists without a cause; and the original cause of this universe (whatever it be) we call GOD." Hume, *Dialogues*, 175.
24 Kemp Smith (1941: 475 n. 1) thinks that this point of Hume's is a recantation of his prior claim, "Of all relations the most universal is that of identity, being common to every being, whose existence has any duration" (T 1.1.5.4; SBN 14). Apparently, Kemp Smith thinks Hume has now denied that the idea identity applies to single beings with duration, contrary to his earlier claim. However, Kemp Smith's point seems to depend on an inaccurate summary of Hume. He paraphrases Hume to be saying, "The view of any one unchanging object is not sufficient to convey the idea of identity" (1941: 474). However, Hume only says, "The view of any one object." He is talking only of single objects represented as such. Hume is not yet talking about steadfast objects represented as such, nor is he talking about steadfast objects fictitiously attributed duration. Once he does, they are precisely what he thinks the idea of identity applies to. After that, any thing that endures, that is, any closely enough related succession, will have the idea of identity applied to it as well. The fiction that all objects that endure have identity is what Hume is referring to at T 1.1.5.4; SBN 14. It is too soon there for him to note that it is a fiction.
25 Bennett 2001: II, 298.
26 George Berkeley, "First Draft of the Introduction to the Principles," 1948–57: Vol. II, 136–7.
27 At the end of this sentence, I retain the word "unity" from the 1739 edition of the *Treatise*, page 351. The first printing of Norton and Norton's Oxford Philosophical Texts edition substitutes "identity." Later printings and the Clarendon edition will return to "unity," and rightly so in my opinion. It seems to me that when Hume next says that the idea we call "identity" is either of "them,"—i.e., either of "unity and number"—"according to the view in which we take it," he needs to have a view in which we take it to be a case of unity. Norton and Norton's initial emendation eliminates that view.
28 For the contrast between coexisting with a succession the way something with duration does, and coexisting with a succession the way a steadfast object does, see Chapter 3, page 45.
29 The key to being a succession, and so having duration, is being several things each occupying distinct moments. Thus, the fact that we think of the distinct

moments and their supposed occupants simultaneously, is not to come to regard them as not successive.

30 Kemp Smith's summary is a bit coarse-grained when he summarizes Hume by saying,

> We both do and do not assert unity; that is to say, we refuse to go the length of number or diversity, and yet restrain ourselves from asserting a strict and absolute unity. Every alleged instance of such identity is an illustration of this self-contradictory procedure; a body is, we believe, both diverse and a unity, a self we believe to be individual and yet also complex, the same with itself and yet in never-ceasing change.
>
> (Kemp Smith 1941: 475)

Hume, since he appeals to switching views of the object, hasn't so baldly committed us to this contradiction when we attribute identity. However, in the end, Kemp Smith is right, as I will show.

31 Why he uses "identity" and "individuation" interchangeably, I don't know.

32 See the first section of Chapter 3, page 33.

33 Bennett does this too at the bottom of Vol. II, p. 298.

34 Green 1886: Sections 255–6, 303.

35 Stroud 1977: 103–4.

36 I explored this possibility in an idiosyncratic way in my dissertation, and my former student Abe Roth developed and extended the suggestion in a sensible, well-argued paper that implicitly draws on the parallels between Hume's difficulty and Frege's puzzle. Baxter 1984: Chapter 2; Roth 1996: 273–98.

37 Cf. T 1.3.1.1; SBN 69: "Two objects, tho' perfectly resembling each other, and even appearing in the same place at different times, may be numerically different."

5 Representing personal identity

1 I will focus on identity. For some discussion of simplicity, see Flage 1991: 147–8.

2 This assumption of accuracy is not an exception to Hume's general skepticism. Hume makes the assumption in the context of taking for granted the philosophical system of the double existence of perceptions and objects. He brings that system into doubt toward the end of T 1.4.2, though returns to it, without answering the doubts, after a bit of "carelessness and in-attention" (T 1.4.2.57; SBN 218). So, as a Pyrrhonian Empiricist, he does not actively endorse the truth of the assumption that consiousness is unerring, yet passively acquiesces in that assumption. See Chapter 1.

3 Locke emphasizes that a person "can consider it self as it self" only by consciousness (2.27.9).

4 It would crowd the mind with perceptions of perceptions to say that each perception is actually reflected in one of consciousness's ideas. The best course, I think, is to follow Johnson's definition of "Conscious": "Endowed with the power of knowing one's own thoughts and actions." So for Hume to say that perceptions are "immediately present to us by consciousness" is to say that we have the power to form ideas of them without the mediation of any other perception or belief (T 1.4.2.47; SBN 212). When coming to have the idea of personal identity, we actually do form these ideas of the train of past perceptions.

I don't yet see that making sense of Hume on consciousness requires the Husserlian account proposed by Waxman (1994) and criticized by Wright as insufficiently supported by the text (Wright 1995: 344–5).

5 In addition to the theories I mention later, Hume may have had others in mind and been sensitive to how they could prevent the direct inference. For some predecessors,

the distinction between distinct perceptions might be a less-than-numerical distinction, such as the modal distinction championed by the great Scholastic metaphysician Suarez, that is, a distinction between substance and mode, or mode and mode. So, an additional premise is needed that the distinction between distinct perceptions is a real distinction, as it was called, a distinction between thing and thing. As Hume puts it, they are distinct existences. That could be why premise (3) is needed. Without it, (6) can be interpreted as not following. Additionally, some philosophers, namely the moderate realists about universals such as Boethius and his heirs, would argue that the mind *can* perceive an identity between distinct existences. From distinct, like particulars the mind can abstract a genuine universal. The likeness between them is just the many distinct things in the world, but is one and the same universal in the mind. Hume rejects this sort of approach with premise (4) that the mind never perceives an identity between distinct existences. Without it, (6) can be interpreted as not a contradiction. So given these further subtleties in the history of metaphysics, Hume may need (3) and (4) to yield the inconsistency that I say he discovers in the Appendix. Suarez 1947: Section 1, pp. 27–37. Boethius 1994: 20–5. I am grateful to William Edward Morris for pressing me on this point.

6 Cf. especially his account of acquiring the idea of body at (T 1.4.2.26–36; SBN 200–5). An example of the sort of thing Hume is talking about can be readily seen by making a flip movie. Put dots in slightly different positions on successive pages of an old book, then riffle the pages. There appears to be a single jittery dot. That dot is really the several dots in succession falsely believed to be identical with each other.

7 Some commentators suggest that, as Bricke puts it, this formulation "will not bear serious attention." (Bricke 1980: 84). However, to understand the problem Hume saw with his account, rather than explore other interesting issues raised by it, we have to take Hume at his word.

8 To look at our perceptions in these different ways is not to "assent to the opinion of a double existence and representation." Because the "generality of mankind" perceive "only one being," they do not think of objects and perceptions as distinct things (T 1.4.2.31; SBN 202).

9 I'm grateful to Andrew Ward and Wade Robison for discussion on this point.

10 See also Patten's argument "that there must be other, non-associative, principles which connect the percepts of a single mind and these he cannot explain" (1976: 60–2).

11 See Bricke's detailed discussion in his Chapter 7, especially 1980: 133–40.

12 *Essay* 2.27.10.

13 As this quotation shows, reflecting ideas are copies of the perceptions they reflect, just like mirror images are copies of what they reflect. (Hume uses "reflection" is the same sense of copy or image at T 1.1.1.3; SBN 2.) Therefore, reflecting on one's past perceptions has nothing to do with what Hume calls "impressions of reflection" and the ideas copied from them. Such impressions are sentiments caused by ideas of our impressions of sensation (T 1.1.2.1; SBN 7–8). Thus, I disagree with a foundational assumption of Stevenson's (1998: 95–129).

14 I'm grateful to Franklin Scott, Thomas Bontly, and Michael Lynch for this objection.

15 Hume gives an explanation of such "confounding" in his discussion of the fictional idea of a vacuum (T 1.2.5.19–21; SBN 60–1).

16 See Stroud 1977: 129–30.

17 "But, as, notwithstanding this distinction and separability, we suppose the whole train of perceptions to be united by identity, a question naturally arises concerning this relation of identity; whether it be something that really binds our several perceptions together, or only associates their ideas in the imagination?" (T 1.4.6.16; SBN 259, my emphasis).

18 Additionally, Hume's third paragraph of the Appendix passage provides a quick rebuttal to the position Flage, rightly I suspect, attributes to Shaftsbury that the self is composed of inseparably connected distinct perceptions. Hume holds if they are distinct then they are separable, which would make Shaftsbury's position inconsistent (T App.12; SBN 634; cf. Flage 1991: 83–7).

19 The fact that both the Theologians and Shaftsbury posit something hidden to disguise the contradiction (cf. T 1.4.6.6; SBN 254–5), does not change the fact that both are committed (according to Hume) to the contradiction of thinking many distinct perceptions are also somehow identical.

20 Cf. Bradley, Chapter 10, "The Reality of the Self" (1897: 89–104).

21 In my previous essay on this matter, I confused the general difficulty with identity that prevents solution to the Appendix problem, and the problem specific to consciousness that appears in the Appendix itself (Baxter 1998).

22 In this way, Hume roughly is concerned with both of the disambiguations of "unite our successive perceptions in our thought or consciousness" which Stroud notes, though I think Stroud is right that Hume's primary concern is with representing the self as identical (Stroud 1977: 133–4). I previously thought that was Hume's only concern.

23 Basson 1958: 131–3; Beauchamp 1979: 37–44; Flage 1991: 149–50; Garrett 1981: 337–58; Haugeland 1998: 68–71; Loeb 1992: 219–31; MacNabb 1951: 149–51; Patten 1976: 59–75; Pears 1993: 289–99; Robison 1976: 79–88; Stroud 1977: 138–9.

24 If Hume, as he appears to, thinks that things outside the mind can exist and yet be nowhere, then the elegant problem Garrett raises is more than a problem with the composition of the self (Garrett 1981: 350–2). Suppose the taste of this olive affects the taste of this pizza, and precisely simultaneously the exactly resembling taste of that olive affects the exactly resembling taste of that pizza. If Garrett is right, Hume cannot say which olive taste affects which pizza taste. Either pairing meets his definitions of cause, or, at least, either does if the *Treatise*'s mention of spatial contiguity is eliminated (T 1.3.14.35; SBN 172). To resolve the problem, Hume must keep to the *Treatise* definition and additionally must posit a non-spatial, but not merely temporal, sort of contiguity that can hold between successive objects. Just as Hume invokes regular spatial contiguity between successive objects to prevent action at a distance, this other sort of contiguity could be invoked to prevent Garrett's analogue to action at a distance (cf. T 1.3.2.6; SBN 75).

25 Johnson 1995: 297–8; Kemp Smith 1941: 73; Passmore 1952: 82–3; Price 1940: 5–6; Robison 1974: 182–93. These other interpretations are applications of Kemp Smith's larger contention that Hume inconsistently holds, under Newtonian influence, that the self is only a bundle of perceptions, and yet, under Hutchesonian influence, that there is a self that operates as an "ever-present *observer*" (Kemp Smith 1941: 73). See also related worries in MacNabb (1951: 151–2) and Penelhum (2000: 57–8).

26 For example, see Pike 1967: 159–65; Penelhum 2000: 54–7; Stroud 1977: 131, and Garrett (1981, 343–5). Presupposing that there must be further explanation is the mistake behind Nathanson's proposal that the mind for Hume is a "set of dispositions" (1976: 40) or a "possessor of propensities" (1976: 45), behind Waxman's unargued claim that association requires "retentive memory" (1992: 236–7), and behind McIntyre's contention that past perceptions have to overlap with present perceptions in order to affect them (McIntyre 1976: 86–8). McIntyre's account is admirable, though, in its attempt to address the concern of the second group of interpretations while staying true to the bundle theory of the first.

27 Garrett 1981: 339–50; Fogelin 1985: 100–5; Penelhum 2000: 48–58. Chapter 3, "Hume's Theory of the Self Revisited," of Penelhum 2000, first appeared under that title in *Dialogue* 14 (1975): 389–409.

28 Kemp Smith 1941: v, 73–6, 171–3, 179–80.
29 Swain (1991: 107) has relied on such interpretations to argue that Hume in the Appendix is not dissatisfied with his theory of personal identity. However, to go on to say his dissatisfaction with his "former opinions" concerns nothing in T 1.4.6, but rather his pre-philosophical opinions that there can be perfect identity through change, seems to me to be stretching the text too far (Swain 1991: 117). For instance, the phrase "binds them together" in Hume's Appendix claim that his account is defective (T App.20; SBN 635), clearly echoes "binds our several perceptions together" at T 1.4.6.16; SBN 259.
30 Ashley and Stack (1974: 239–54; Biro 1976: 19–38; Kemp Smith 1941: 96–8, 499–502; Laird 1931: 171–2; MacNabb 1951: 147–8, 151; Penelhum 2000: Chapters 2 and 3; Perry 1975: 26–30. Chapter 2, "Hume on Personal Identity," of Penelhum 2000, first appeared under that title in *Philosophical Review* 64 (1955): 571–89.
31 The only way I know of to make sense of the identity of distinct things is the theory I propose in "Many-One Identity," and "Identity in the Loose and Popular Sense," but such a theory cannot in good conscience be attributed to Hume or to these commentators.
32 It is "identity, in an improper sense" and a "mistake" (T 1.4.6.7; SBN 255). Cf. T 1.4.6.9; SBN 256 and T 1.4.6.16; SBN 260. As Reid says, imperfect identity holds between two objects which cannot be the same, but which we are apt to think of as the same. It is not identity but something that "for the conveniency of speech, we call identity" (Hamilton 1872: 346). See also Kornegay 1985: 213–26.
33 Hume's account in this respect is just like his account of getting the idea of body: "[W]e are not apt to regard these interrupted perceptions as different, (which they really are) but on the contrary consider them as individually the same" (T 1.4.2.24; SBN 199).
34 This fact is reinforced by the fact that Hume is also trying to explain the attribution of *simplicity* to what is really complex (T App; SBN 635). He is not giving "simplicity conditions" for bundles.
35 See Chapter 2. Penelhum 2000: 30–2.
36 Bricke 1980: 88, 159 n. 13; Ainslie 2001: 566; Penelhum 2000: 117.
37 Thus I think Penelhum is right to hesitate about Ainslie's and his view (2000: 119).
38 Pitson 2002: 76–8.
39 Pitson 2002: 71–5. I observe that Hume did have available a constancy among perceptions considered as perceptions in the mind to supplement any lack of coherence: "that *je-ne-sçai-quoi*, of which 'tis impossible to give any definition or description, but which every one sufficiently understands" (T 1.3.8.16; SBN 106). I must grant to Pitson, however, that Hume does not appeal to this constancy.
40 Pitson 2002: 74.
41 For example Fogelin (1985: 100) lends support to the first, and (1985: 105, 108) to the second. He forthrightly admits that his own suggestive proposals have trouble meeting the second criterion. Stroud (1977: 128) lends support to the third. Ainslie breaks the second criterion into several more specific ones (2001: 567–77). Pitson endorses these and argues convincingly for the third (2002: 66, 174 n. 11). See also Beauchamp 1979: 38, 41.
42 Don Garrett emphasized this criterion as I was writing "Hume's Labyrinth." Pitson's book forced me to revisit it, beginning the process that transformed that former article into this almost entirely different chapter.
43 I'm grateful to Terrence Penelhum for pressing me on this point.
44 Penelhum 2000: 103.
45 Cf. Descartes's goal of establishing something "in the sciences that was stable and likely to last" (Descartes 1984: 12).
46 I am grateful for suggestions or criticism from Mark Rubin, Elise Springer, Eddie Zemach, Jerome Shaffer, Phillip Cummins, Wade Robison, Martha Nussbaum,

James van Cleve, John Wright, Sandra Baxter, Ted Morris, Don Garrett, Lionel Shapiro, Franklin Scott, Michael Lynch, Thomas Bontly, and some anonymous referees.

6 Systematic exposition of Hume's difficulty

1 Frege 1953: 56–7. Here I give just the original version of Frege's puzzle, which Salmon has shown is really a general puzzle about substitutivity. See my Chapter 4, note 2.
2 I am leaving it open what the existence of intentional objects consists in. The relation between intentional objects and intended objects is important, but not my concern here.
3 Berkeley, *Principles*, Part I, Section 51 (1948–57: II, 62–3).
4 Clark suggests that color ascription is an eight-term relation (2000: Chapter 6). Despite the extraordinary difficulty in making sense of what it is for a color ascription to be correct, I'm assuming we can make sense of it well enough to serve the purposes of, say, a kindergarten teacher teaching children their colors.
5 See T 1.3.1.1; SBN 69. Hume thought only statements of identity through time were like this, but I am neglecting this part of his account. See T 1.1.5.4, 1.4.2.29–30; SBN 14, 200–1.
6 I will say "for all we know" though sometimes it is more appropriate to say "for all we knew". As a variant of "for all we know" I will say "perhaps".
7 I am assuming that we can represent there as being absurdities, whether or not they are clear to us.
8 This alternate consideration is Hume's own "solution" to the problem. But Hume is just concerned to give a causal explanation of how we come to overlook the problem; his concern is not really to resolve it. See T 1.4.2.29; SBN 200–1.
9 The seeming ungrammaticality of "is the same thing(s)" is a reflection of Hume's difficulty.
10 An appeal to Geach's relative identity, that is identity relative to sorts, won't help. The same sortal would apply both to the things as identical and as distinct (Geach 1967).
11 I do not need to assume that such a sequence can completely capture what one represents there as being, nor that only one sequence of sentences is the correct one.
12 For antecedent appeals to stories see Adams 1983: 218, and Camp 1988: 11–14.
13 Think of "p" as a dummy constant to be replaced by sentences.
14 It is true, of course, that we can imagine ourselves knowing things beyond what we know. But such alternate stories are not alternate ways of representing how things really are beyond what is actually known. This latter is the concern here. Thus the relation between my alternate stories should not be confused with Hintikka's relation between epistemic alternatives. See Hintikka 1962: 45.
15 This characterization of alternates could be generalized by saying one story is an alternate to another relative to which cast of characters and which characterizations are kept constant. An alternate represents there as being a situation that in a given story is perhaps true, just in case the characters and characterizations kept constant are determined by the relevant things we know.
16 Clumsily, but more perspicuously, this might be put: In Story B, Cicero is such that perhaps something identical with him is distinct from Tully. The same treatment of (8) might ease the transition to (9).
17 Kripke's famous claim about possible worlds certainly applies *mutatis mutandis* to alternate stories: "'Possible worlds' are *stipulated*, not *discovered* by powerful telescopes. There is no reason why we cannot *stipulate* that, in talking about what would have happened to Nixon in a certain counterfactual situation, we are

talking about what would have happened to *him*" (Kripke 1980: 44). Thus, alternate stories B and C are not to be confused with Kripkean "qualitatively identical epistemic situations" such that something in one is called by the same names as distinct things in the other because exactly resembling evidence is presented in both (Kripke 1980: 104). Likewise, alternate stories B and C are not to be confused with Hintikkan "epistemic alternatives" such that something in one is called by the same names as distinct things in the other because the knower does not know the ways the names in fact refer (Hintikka 1962: 138–41).

18 We can, of course, tell a story according to which some other story is consistent or true. However, that only makes it consistent or true according to a story, not in fact. That storytellers are concerned to correct inadvertent inconsistencies, shows that they recognize that there can in fact be such.

19 This does not prove that the puzzles are irrelevant to each other. A solution to Hume's might give those who abjure senses additional resources for addressing Frege's.

20 This fact rules out an alternate approach to resolving the difficulty that could be inspired by Lewis's counterpart theory: Rule that no two stories ever have the same characters, and that what is perhaps true of a character in one story is a matter of what is true of an appropriately resembling character in another (Lewis 1968).

21 I'm grateful for helpful comments and criticism from John Troyer, Sam Wheeler, Austen Clark, Scott Lehmann, Tim Elder, Steve Lahey, Elise Springer, and especially an anonymous referee for *Philosophical Studies.*

Conclusion

1 I've made some halting steps in the direction of a solution in "Many-One Identity" (Baxter 1988b); "Identity in the Loose and Popular Sense" (Baxter 1988c); "The Discernibility of Identicals" (Baxter 1999); "Instantiation as Partial Identity" (Baxter 2001c); and "Altruism, Grief, and Identity" (Baxter 2005).

2 At T 1.4.6.6; SBN 254, Hume talks also of interrupted successions, but that is because he is there talking of the attribution of identity to plants and animals, etc., as well as to the self.

3 Pitson (2002: 30) suggests that the vulgar only believe in some "unknown and mysterious" connection, as opposed to a soul, however it seems to me that Hume uses that phrase in connection with "plants and vegetables" (T 1.4.6.6; SBN 255).

4 Talk of the mind taking and supplying is metaphorical for an account purely in terms of successive perceptions. See Pike's article (1967). Eventually, in the case of body, identity is attributed in some cases even when perceptions are both "interrupted and variable" (T 1.4.6.6; SBN 254, see also T 1.4.6.14; SBN 258). Pitson suggests that there can be interruption even in the case of the self, while acknowledging that there can be no experienced interruption (2002: 74–75).

5 Also Hume says, "For as such a succession answers evidently to our notion of diversity, it can only be by a mistake we ascribe to it an identity" (T 1.4.6.7; SBN 255).

6 Peirce 1955: 14–18.

7 For a recent defense of "neo-Pyrrhonian" skepticism, see Fogelin 1994.

Bibliography

Adams, Robert M. (1983) "Phenomenalism and Corporeal Substance in Leibniz," in Peter A. French, Theodore Edward Uehling, and Howard K. Wettstein (eds) *Midwest Studies in Philosophy.* Minneapolis, Minn.: University of Minnesota Press, pp. 217–58.

Ainslie, Donald C. (2001) "Hume's Reflections on the Identity and Simplicity of Mind," *Philosophy and Phenomenological Research* 62: 557–78.

Annas, Julia (2000) "Hume and Ancient Scepticism," *Acta Philosophica Fennica* 66: 271–85.

—— (ed.) (2001) *Voices of Ancient Philosophy.* New York: Oxford University Press.

Aristotle (1941) *The Basic Works of Aristotle.* Edited by Richard McKeon. New York: Random House.

Armstrong, David M. (1997) *A World of States of Affairs.* Cambridge: Cambridge University Press.

Arnauld, Antoine, and Pierre Nicole (1996) *Logic or the Art of Thinking.* Translated by Jill Vance Buroker. New York: Cambridge University Press.

Ashley, Lawrence and M. Stack (1974) "Hume's Theory of the Self and Its Identity," *Dialogue* 13: 239–54.

Augustine (1961) *Confessions.* Translated by R. S. Pine-Coffin. New York: Penguin Books.

Ayers, Michael (2003) "Ideas and Objective Being," in Daniel Garber and Michaels Ayers (eds) with the assistance of Roger Ariew and Alan Gabbey, *The Cambridge History of Seventeenth-Century Philosophy.* Cambridge: Cambridge University Press, Vol. II, pp. 1062–107.

Baier, Annette C. (1991) *A Progress of Sentiments: Reflections on Hume's* Treatise. Cambridge, Mass.: Harvard University Press.

Basson, Anthony H. (1958) *David Hume.* London: Penguin.

Baxter, Donald L. M. (1984) "The One and the Many: Developing Hume's Account of Identity," Ph.D. dissertation, University of Pittsburgh.

—— (1987) "A Defense of Hume on Identity through Time," *Hume Studies* 13: 323–42.

—— (1988a) "Hume on Infinite Divisibility," *History of Philosophy Quarterly* 5: 133–40.

—— (1988b) "Many-One Identity," *Philosophical Papers* 17: 193–216.

—— (1988c) "Identity in the Loose and Popular Sense," *Mind* 97: 575–82.

—— (1990) "Hume on Virtue, Beauty, Composites, and Secondary Qualities," *Pacific Philosophical Quarterly* 71: 103–18.

—— (1995) "Corporeal Substances and True Unities," *Studia Leibnitiana* 27: 157–84.

—— (1997) "Abstraction, Inseparability, and Identity," *Philosophy and Phenomenological Research* 57: 307–30.

—— (1998) "Hume's Labyrinth Concerning the Idea of Personal Identity," *Hume Studies* 24: 203–33.

—— (1999) "The Discernibility of Identicals," *Journal of Philosophical Research* 24: 37–55.

——(2000a) "A Humean Temporal Logic," in Akihiro Kanamori (ed.) *The Proceedings of the Twentieth World Congress of Philosophy*, Volume 6: *Analytic Philosophy and Logic*. Bowling Green, Ohio: Philosophy Documentation Center, pp. 209–16.

—— (2000b) "Hume's Puzzle About Identity," *Philosophical Studies* 98: 187–201.

—— (2001a) "Loose Identity and Becoming Something Else," *Noûs* 35: 592–601.

—— (2001b) "Hume on Steadfast Objects and Time," *Hume Studies* 27: 129–48.

—— (2001c) "Instantiation as Partial Identity," *Australasian Journal of Philosophy* 79: 449–64.

—— (2005) "Altruism, Grief, and Identity," *Philosophy and Phenomenological Research* 70: 371–83.

—— (forthcoming) "Hume's Theory of Space and Time in Its Sceptical Context," in David Fate Norton (ed.) *The Cambridge Companion to Hume*, 2nd edn. Cambridge: Cambridge University Press.

Bayle, Pierre (1991) *Historical and Critical Dictionary: Selections*. Translated by Richard H. Popkin. Indianapolis, Ind.: Hackett.

Beauchamp, Tom L. (1979) "Self Inconsistency or Mere Self Perplexity?" *Hume Studies* 5: 37–44.

Bennett, Jonathan (2001) *Learning from Six Philosophers: Descartes, Spinoza, Leibniz, Locke, Berkeley, Hume*. 2 vols. Oxford: Clarendon Press.

Berkeley, George (1948–57) *The Works of George Berkeley, Bishop of Cloyne*. Edited by A. A. Luce and T. E. Jessop. 9 vols. London: Thomas Nelson.

—— (1982) *A Treatise Concerning the Principles of Human Knowledge*. Edited by Kenneth Winkler. Indianapolis, Ind.: Hackett.

Biro, John I. (1976) "Hume on Self-Identity and Memory," *Review of Metaphysics* 30: 19–38.

Boethius (1994) "Isagoge," in Paul Vincent Spade (ed.) *Five Texts on the Mediaeval Problem of Universals*. Indianapolis, Ind.: Hackett, pp. 20–5.

Boler, John (1963) *Charles Pierce and Scholastic Realism*. Seattle, Wash.: University of Washington Press.

Bracken, Harry M. (1984) "Hume on the 'Distinction of Reason'," *Hume Studies* 10: 81–108.

Bradley, F. H. (1897) *Appearance and Reality*. 2nd edn. Oxford: Clarendon Press.

Broughton, Janet (2004) "The Inquiry in Hume's *Treatise*," *The Philosophical Review* 113: 537–56.

Brentano, Franz (1995) *Psychology from an Empirical Standpoint*. Edited by Linda L. McAlister. London: Routledge.

Bricke, John (1980) *Hume's Philosophy of Mind*. Princeton, NJ: Princeton University Press.

Broad, C. D. (1961) "Hume's Doctrine of Space," *Proceedings of the British Academy* 47: 161–76.

Burnyeat, Myles (ed.) (1983) *The Skeptical Tradition*. Berkeley, Calif.: University of California Press.

Butler, Joseph (1849) *The Analogy of Religion*. Edited by Samuel Halifax. Oxford: Oxford University Press.

Butler, Ronald J. (1975) "'Distinctiones Rationis' or the Cheshire Cat Which Left Its Smile Behind It," *Proceedings of the Aristotelian Society* 76: 165–76.

Butts, Robert E. (1959) "Husserl's Critique of Hume's Notion of 'Distinctions of Reason'," *Philosophy and Phenomenological Research* 20: 213–21.

Camp, Joseph L., Jr. (1988) "Why Attributes of Aboutness Report Soft Facts," *Philosophical Topics* 16: 5–30.

Carey, Susan, and Fei Xu (2001) "Infants' Knowledge of Objects: Beyond Object Files and Object Tracking," *Cognition* 80: 179–213.

Carnap, Rudolf (1967) "On the Character of Philosophic Problems," in Richard Rorty (ed.) *The Linguistic Turn*. Chicago, Ill.: University of Chicago Press, pp. 54–62.

Cicero (2006) *On Academic Scepticism*. Translated by Charles Brittain. Cambridge and Indianapolis, Ind.: Hackett Publishing.

Clark, Austen (2000) *A Theory of Sentience*. Oxford: Oxford University Press.

Costa, Michael J. (1990) "Hume, Strict Identity, and Time's Vacuum," *Hume Studies* 16: 1–16.

Cummins, Phillip (1990) "Bayle, Leibniz, Hume and Reid on Extension, Composites and Simples," *History of Philosophy Quarterly* 7: 299–314.

Davidson, Donald (2001) *Inquiries into Truth and Interpretation*. 2nd edn. Oxford: Oxford University Press.

Descartes, René (1984) *Meditations on First Philosophy*. Translated by John Cottingham. In *The Philosophical Writings of Descartes*, Vol. II. Translated by John Cottingham, Robert Stoothoff, and Dugald Murdoch. Cambridge: Cambridge University Press.

Falkenstein, Lorne (1997) "Hume on Manners of Disposition and the Ideas of Space and Time," *Archiv für Geschichte der Philosophie* 79: 179–201.

Flage, Daniel E. (1986) "Berkeley on Abstraction," *Journal of the History of Philosophy* 24: 483–501.

—— (1991) *David Hume's Theory of Mind*. New York: Routledge.

Flew, Antony (1976) "Infinite Divisibility in Hume's Treatise," in Donald W. Livingston and James T. King (eds) *Hume: A Re-Evaluation*. New York: Fordham University Press, pp. 257–69.

Fodor, Jerry A. (2003) *Hume Variations*. Oxford: Clarendon Press.

Fogelin, Robert (1985) *Hume's Skepticism in the Treatise of Human Nature*. London: Routledge & Kegan Paul.

—— (1988) "Hume and Berkeley on the Proofs of Infinite Divisibility," *Philosophical Review* 97: 47–69.

—— (1994) *Pyrrhonian Reflections on Knowledge and Justification*. New York: Oxford University Press.

Fosl, Peter S. (1998) "The Bibliographic Bases of Hume's Understanding of Sextus Empiricus and Pyrrhonism," *Journal of the History of Philosophy* 36: 261–78.

Franklin, James (2001) *The Science of Conjecture: Evidence and Probability before Pascal*. Baltimore, Md. and London: Johns Hopkins University Press.

Frasca-Spada, Marina (1998) *Space and the Self in Hume's Treatise*. Cambridge: Cambridge University Press.

Frede, Michael (1987) "The Skeptic's Two Kinds of Assent and the Question of the Possibility of Knowledge," in *Essays in Ancient Philosophy*. Minneapolis, Minn.: University of Minnesota Press, pp. 201–22.

Frege, Gottlob (1953) *The Foundations of Arithmetic: A Logico-Mathematical Enquiry into the Concept of Number*. Translated by J. L. Austin. 2nd edn. New York: Harper Brothers.

—— (1980) "On Sense and Meaning," in Max Black and Peter Geach (eds) *Translations from the Philosophical Writings of Gottlob Frege*. Oxford: Basil Blackwell, pp. 56–78.

Garrett, Don (1981) "Hume's Self-Doubts About Personal Identity," *Philosophical Review* 90: 337–58.

—— (1997) *Cognition and Commitment in Hume's Philosophy*. Oxford: Oxford University Press.

Geach, Peter (1967) "Identity," *The Review of Metaphysics* 21: 3–12.

Grajewski, Maurice J. (1944) *The Formal Distinction of Duns Scotus*. Washington, DC: Catholic University of America Press.

Green, T. H. (1886) "General Introduction," in David Hume, *A Treatise of Human Nature*, Vol. I, edited by T. H. Green and T. H. Grose. London: Longmans, Green, & Co.

Grünbaum, Adolf (1967) *Modern Science and Zeno's Paradoxes*. London: Allen & Unwin.

Hamilton, William (ed.) (1872) *The Works of Thomas Reid*. 7th edn, 2 vols. Edinburgh: MacLachlan & Stewart.

Haugeland, John (1998) "Hume on Personal Identity," in *Having Thought: Essays in the Metaphysics of Mind*. Cambridge, Mass.: Harvard University Press, pp. 63–71.

Hempel, Carl G. (1959) "The Empiricist Criterion of Meaning," in A. J. Ayer (ed.) *Logical Positivism*. New York: The Free Press.

Hintikka, Jaakko (1962) *Knowledge and Belief*. Ithaca, NY: Cornell University Press.

Hirsch, Eli (1983) "Hume's Distinction between Genuine and Fictitious Identity," *Midwest Studies in Philosophy* 8: 321–38.

Holden, Thomas (2002) "Infinite Divisibility and Actual Parts in Hume's Treatise," *Hume Studies* 28 : 3–25.

—— (2004) *The Architecture of Matter: Galileo to Kant*. Oxford: Clarendon Press.

Hume, David (1739) *A Treatise of Human Nature: Being an Attempt to Introduce the Experimental Method of Reasoning into Moral Subjects*, Vol. I. London: Printed for John Noon.

—— (1886) *A Treatise of Human Nature*. Edited by T. H. Green and T. H. Grose. 2 vols. London: Longmans, Green, & Co.

—— (1935) *Dialogues Concerning Natural Religion*. Edited by Norman Kemp Smith. Oxford: Clarendon Press.

—— (1975) *Enquiries Concerning Human Understanding and Concerning the Principles of Morals*, 3rd edn. Oxford: Clarendon Press.

—— (1978) *A Treatise of Human Nature*. Edited by L. A. Selby-Bigge and P. H. Nidditch. 2nd edn. Oxford: Clarendon Press.

—— (1983) *History of England, from the Invasion of Julius Caesar to the Revolution in 1688*. 6 vols. Indianapolis, Ind.: Liberty Classics.

—— (1985) "Of the Standard of Taste," in Eugene F. Miller (ed.) *Essays Moral, Political, and Literary*. Indianapolis, Ind.: Liberty Fund, pp. 226–49.

—— (2000a) *A Treatise of Human Nature*. Edited by David Fate Norton and Mary J. Norton. Oxford: Oxford University Press.

—— (2000b) *An Enquiry Concerning Human Understanding*. Edited by Tom L. Beauchamp. Oxford: Clarendon Press.

Johnson, Oliver A. (1995) *The Mind of David Hume.* Champaign, Ill.: University of Illinois Press.

Johnson, Samuel (1755) *A Dictionary of the English Language.* London: W. Strahan.

Kanamori, Akihiro (ed.) (2000) "Introduction," *The Proceedings of the Twentieth World Congress of Philosophy,* Vol. VI: *Analytic Philosophy and Logic.* Bowling Green, Ohio: Philosophy Documentation Center, pp. xiii–xli.

Kanizsa, Gaetano (1987) "Quasi-Perceptual Margins in Homogeneously Stimulated Fields," in Susan Petry and Glenn E. Meyer (eds) *The Perception of Illusory Contours.* Translated by Walter Gerbino. New York: Springer-Verlag.

Kemp Smith, Norman (1941) *The Philosophy of David Hume.* London: Macmillan.

Kornegay, R. Jo (1985) "Hume on Identity and Imperfect Identity," *Dialogue* 24: 213–26.

Kripke, Saul A. (1980) *Naming and Necessity.* Cambridge, Mass.: Harvard University Press.

Laird, John (1931) *Hume's Philosophy of Human Nature.* New York: Dutton.

Leibniz, Gottfried (1967) *The Leibniz–Arnauld Correspondence.* Translated by H. T. Mason. Manchester: Manchester University Press.

Lewis, David (1968) "Counterpart Theory and Quantified Modal Logic," *Journal of Philosophy* 65: 113–26.

—— (1986) *On the Plurality of Worlds.* Oxford: Blackwell.

—— (1991) *Parts of Classes.* Oxford: Basil Blackwell.

Locke, John (1975) *An Essay Concerning Human Understanding.* Edited by Peter H. Nidditch. Oxford: Clarendon Press.

Loeb, Louis (1992) "Causation, Extrinsic Relations, and Hume's Second Thoughts About Personal Identity," *Hume Studies* 18: 219–31.

—— (2002) *Stability and Justification in Hume's* Treatise. Oxford: Oxford University Press.

MacNabb, D. G. C. (1951) *David Hume: His Theory of Knowledge and Morality.* London: Hutchinson.

Mandelbaum, Maurice (1974) "The Distinguishable and the Separable: A Note on Hume and Causation," *Journal of the History of Philosophy* 12: 242–7.

Mates, Benson (ed.) (1996) *The Skeptic Way: Sextus Empiricus's* Outlines of Pyrrhonism. Oxford: Oxford University Press.

McIntyre, Jane L. (1976) "Is Hume's Self Consistent?" in David Fate Norton, Nicholas Capaldi and Wade L. Robison (eds) *McGill Hume Studies.* San Diego, Calif.: Austin Hill Press, pp. 79–88.

McRae, Robert (1980) "The Import of Hume's Theory of Time," *Hume Studies* 6: 119–32.

Mossner, Ernest (1980) *The Life of David Hume,* 2nd edn. Oxford: Clarendon Press.

Nathanson, Stephen (1976) "Hume's Second Thoughts on the Self," *Hume Studies* 2: 36–45.

Newman, Rosemary (1995) "Hume on Space and Geometry," in Stanley Tweyman (ed.) *David Hume: Critical Assessments.* London: Routledge & Kegan Paul, pp. 39–60.

Newton, Isaac (1962) *Mathematical Principles of Natural Philosophy.* Translated by Florian Cajori. Berkeley and Los Angeles, Calif.: University of California Press.

Norton, David Fate, Nicholas Capaldi, and Wade L. Robison (eds) (1976) *McGill Hume Studies.* San Diego, Calif.: Austin Hill Press.

Norton, David Fate (1982) *David Hume: Common-Sense Moralist, Sceptical Metaphysician.* Princeton, NJ: Princeton University Press.

Passmore, John A. (1952) *Hume's Intentions.* Cambridge: Cambridge University Press.

Patten, Steven C. (1976) "Hume's Bundles, Self-Consciousness and Kant," *Hume Studies* 2: 59–75.

Pears, David (1993) "Hume on Personal Identity," *Hume Studies* 19: 289–99.

Peirce, Charles Sanders (1955) "The Fixation of Belief," in Justus Buchler (ed.) *Philosophical Writings of Peirce.* New York: Dover Publications. pp. 5–22.

Penelhum, Terence (1955) "Hume on Personal Identity," *Philosophical Review* 64: 571–89.

—— (1975) "Hume's Theory of the Self Revisited," *Dialogue* 14: 389–409.

—— (2000) *Themes in Hume: The Self, the Will, Religion.* Oxford: Clarendon Press.

Perry, John (ed.) (1975) *Personal Identity.* Berkeley, Calif.: University of California Press.

Pike, Nelson (1967) "Hume's Bundle Theory of the Self: A Limited Defense," *American Philosophical Quarterly* 4: 159–65.

Pitson, A. E. (2002) *Hume's Philosophy of the Self.* London and New York: Routledge.

Plutarch (2001) "On Common Conceptions," 1081c–82a in Julia Annas (ed.) *Voices of Ancient Philosophy: An Introductory Reader.* New York: Oxford University Press, p. 286.

Popkin, Richard H. (1966) "David Hume: His Pyrrhonism and His Critique of Pyrrhonism," in V. C. Chappell (ed.) *Hume: A Collection of Critical Essays.* Garden City, NY: Doubleday, pp. 53–98.

—— (2003) *The History of Scepticism from Savonarola to Bayle.* Oxford: Oxford University Press.

Price, Henry H. (1940) *Hume's Theory of the External World.* Oxford: Clarendon Press.

Pylyshyn, Zenon W. (2001) "Visual Indexes, Preconceptual Objects, and Situated Vision," *Cognition* 80: 127–58.

Quine, W. V. (1956) "Quantifiers and Propositional Attitudes," *Journal of Philosophy* 53: 177–87.

Read, R. and Richman, K. A. (eds) (2000) *The New Hume Debate.* London and New York: Routledge.

Robison, Wade L. (1974) "Hume on Personal Identity," *Journal of the History of Philosophy* 12: 182–93.

—— (1976) "In Defense of Hume's Appendix," in David Fate Norton, Nicholas Capaldi and Wade L. Robison (eds) *McGill Hume Studies.* San Diego, Calif.: Austin Hill Press, pp. 79–88.

Rosenberg, Alexander (1993) "Hume and the Philosophy of Science," in David Fate Norton (ed.) *The Cambridge Companion to Hume.* New York: Cambridge University Press, pp. 64–89.

Roth, Abraham Sesshu (1996) "Hume's Psychology of Identity Ascriptions," *Hume Studies* 22: 273–98.

Russell, Bertrand (1985) *The Philosophy of Logical Atomism.* Edited by David Pears. Chicago, Ill.: Open Court.

Salmon, Nathan (1986) *Frege's Puzzle.* Cambridge, Mass.: MIT Press.

Sellars, Wilfred (1979) *Naturalism and Ontology.* Reseda, Calif.: Ridgeview.

Sextus Empiricus (2000) *Outlines of Scepticism.* Edited by Julia Annas and Jonathan Barnes. Cambridge: Cambridge University Press.

Spade, Paul Vincent (ed.) (1994) *Five Texts on the Mediaeval Problem of Universals: Porphyry, Boethius, Abelard, Duns Scotus, Ockham.* Indianapolis, Ind.: Hackett.

Stevenson, Gordon Park (1998) "Humean Self-Consciousness Explained," *Hume Studies* 24: 95–129.

Stroud, Barry (1977) *Hume*. London: Routledge & Kegan Paul.

Suarez, Francis (1947) *On the Various Kinds of Distinctions (Disputationes Metaphysicae. Disputatio VII, de variis distinctionum generibus)*. Translated by Cyril Vollert. Milwaukee, Wisc.: Marquette University Press.

Swain, Corliss Gayda (1991) "Being Sure of One's Self: Hume on Personal Identity," *Hume Studies* 17: 107–24.

Traiger, Saul (1987) "Hume on Impressions, Ideas, and Fictions," *Hume Studies* 13: 381–99.

Tweyman, Stanley (1974) "Hume on Separating the Inseparable," in William B. Todd (ed.) *Hume and the Enlightenment*. Edinburgh: Edinburgh University Press, pp. 30–42.

Van Leeuwen, Henry G. (1963) *The Problem of Certainty in English Thought: 1630–1690*. The Hague: Martinus Nijhoff.

van Steenburgh, E. W. (1977) "Durationless Moments in Hume's Treatise," in George P. Morice (ed.) *David Hume: Bicentenary Papers*, Austin, Tex.: University of Texas Press, pp. 181–5.

Waxman, Wayne (1992) "Hume's Quandary Concerning Personal Identity," *Hume Studies* 18: 233–53.

—— (1994) *Hume's Theory of Consciousness*. New York: Cambridge University Press.

Weinberg, Julius R. (1965) *Abstraction, Relation and Induction: Three Essays in the History of Thought*. Madison, Wisc.: University of Wisconsin Press.

Williams, Donald C. (1951) "The Myth of Passage," *The Journal of Philosophy* 48: 457–72.

Winkler, Kenneth (1983) "Berkeley on Abstract Ideas," *Archiv für Geschichte der Philosophie* 65: 69–73.

—— (1989) *Berkeley: An Interpretation*. New York: Clarendon Oxford Press.

Wittgenstein, Ludwig (1961) *Tractatus Logico-Philosophicus*. London: Routledge & Kegan Paul.

Wright, John P. (1983) *The Sceptical Realism of David Hume*. Minneapolis, Minn.: University of Minnesota Press.

—— (1995) "Critical Study of Wayne Waxman's Hume's Theory of Consiousness," *Hume Studies* 21: 344–50.

Index

For Product Safety Concerns and Information please contact our EU representative GPSR@taylorandfrancis.com
Taylor & Francis Verlag GmbH, Kaufingerstraße 24, 80331 München, Germany

www.ingramcontent.com/pod-product-compliance
Ingram Content Group UK Ltd.
Pitfield, Milton Keynes, MK11 3LW, UK
UKHW011019180926
475701UK00004B/838

* 9 7 8 0 4 1 5 8 0 4 7 7 6 *